What the Experts Say About Healthy Exchanges®

"I have found the recipes directed at low fat options to be a great future resource for women seeking alternatives to oral calcium supplements. Many women view traditional calcium-rich food sources as having a high fat content. Additionally, many women are wanting to make healthy choices through their diet versus adding another "pill" to their daily regimen. The recipes will also provide an option for women who have not tolerated taking an oral calcium supplement due to the common side effects of gastric upset and increased intestinal gas (flatulence)."

> —Rebecca Jacobs, R.N.C., M.A.,
> Clinical Nurse Specialist, Iowa Women's Health Center,
> Children and Women's Services,
> Iowa City, IA

"JoAnna's recipes provide an excellent way to incorporate more calcium in your diet while maintaining a low fat approach to eating. Many women find it difficult to get an adequate amount of calcium daily. JoAnna provides a fun, easy and tasty way to accomplish this goal."

> —Kerry Humes, M.D.,
> Women's Health Center,
> Moline, IL

"A large number of my patients consume one or more caffeine beverages a day, increasing their risk for osteoporosis, and drink very little milk. JoAnna's quick, simple and tasty recipes will enable women to increase their calcium intake."

> —Anne G. Rapp, R.N., B.S.N., A.R.N.P.,
> Premier Care P.C.,
> Davenport, IA

"I highly recommend *The Strong Bones Healthy Exchanges Cookbook* to my patients as a nutritious and delicious way to increase their daily calcium intake. I especially enjoy the 'Midwestern appeal' of the recipes as well as the nutrition analysis and ADA food exchanges, which are provided for each recipe."

> —Becky Goodell, R.D., L.D.,
> Director of Out-Patient Nutrition Services,
> Albert Lea Medical Center,
> Albert Lea, MN

"An often-heard request from my clients is for 'normal' recipes that are healthy and fit the dietary needs of the individuals. How wonderful to come across a cookbook author who identifies the real concerns of the average person: time, taste, ease of preparation and use of common ingredients."

> —Kathy Reinke, R.D., C.D., C.D.E.,
> Shawano Medical Center
> Shawano, WI

The

Strong Bones

Healthy

Exchanges®

Cookbook

JoAnna M. Lund

Introduction by
Brian L. Levy, M.D., F.A.C.E.

A Perigee Book

A Perigee Book
Published by The Berkley Publishing Group
200 Madison Avenue
New York, NY 10016
A member of Penguin Putnam Inc.

For more information about Healthy Exchanges products, contact:
   Healthy Exchanges, Inc.
   PO Box 124
   DeWitt, IA 52742-0124
   (319) 659-8234

First edition: September 1997

Published simultaneously in Canada.

The Putnam Berkley World Wide Web site address is http://www.berkley.com

Library of Congress Cataloging-in-Publication Data

Lund, JoAnna M.
   The strong bones healthy exchanges cookbook / JoAnna M. Lund;
introduction by Brian L. Levy.
      p.   cm.
   "A Perigee book."
   Includes index.
   ISBN 0-399-52337-5
   1. High-calcium diet—Recipes.
RM237.56.L86 1997
641.5'632—dc21   97-3466
                                                         CIP

Printed in the United States of America

10   9   8   7   6   5   4   3   2   1

As all my books are, this cookbook is dedicated in loving memory to my parents, Jerome and Agnes McAndrews, who taught me the virtue of "making do" with what you have. I inherited from them the talent for "making do" with dry milk powder while creating healthy recipes rich in calcium *and* taste.

This cookbook is also dedicated to the dairy farmers of America. Without their hard work and long hours, we couldn't so easily enjoy both the convenience and the calcium-rich foods that dairy products provide. The next time you reach for a glass of ice-cold milk, please thank the men and women who rise before dawn to ensure milk and other dairy products are readily available.

As usual, my mother had the perfect poem for this book in her vast collection. While she was writing about the loss of Autumn, the analogy is ideal for the loss of calcium from our bones which causes osteoporosis, a condition in which so many suffer the consequences of calcium deficiency before they know they have a problem.

Jack Frost

Jack Frost paid Autumn a visit
    while she was packing to go.
He warned her that she must leave
    before the falling snow.
Then he began to strip the bright trees
    and left the branches bare.
Exposing their shivering, naked limbs
    to the crisp and frosty air.
Like an avalanche of confetti,
    leaves tumbled to the ground
As Fall quickly grabbed her paint cans
    and left without a sound.

—*Agnes Carrington McAndrews*

Contents

Acknowledgments

I love stirring up "common folk" healthy recipes as much today as when I started over five years ago. And I love helping others help themselves with those recipes. My goal for this collection of recipes is to help current and future generations avoid paying the price for not consuming enough calcium-rich foods. For helping me to try for that goal, I want to thank:

John Duff, my editor and cheerleader. He continues to encourage me to create.

Angela Miller and Coleen O'Shea, my agents. A better dynamic duo can't be found.

Shirley Morrow and Wendi Seatrand for typing, typing, typing, and then typing all over again. They learned firsthand that I can neither spell nor write legibly.

Rita Ahlers and Gerry Stamp for helping me test my recipes. They willingly fit their schedules into mine so we could get the job done on time.

Lori Hansen for calculating the recipes so the nutrient information could be as current as the products in our stores. She is a master when it comes to the keyboard.

Rose Hoenig, R.D., L.D., for answering my nutritional questions in "real world" language and for calculating the Diabetic Exchanges. She is a professional in every sense of the word.

Brian L. Levy, M.D., F.A.C.E., for writing the introduction. He made it easy for the rest of us to understand why a calcium-rich diet is so important for so many reasons.

Barbara Alpert for helping me gather my thoughts in a coherent way. Not only has she become a friend, she has helped me improve my writing while still retaining my "Grandma Moses" style.

Cliff Lund, my "Official Taste Tester." He still tests my recipes with a smile on his face and a kiss on his lips.

God, for giving me the ability to create. Through God all things are possible, including easy, tasty, and healthy recipes.

Introduction

No matter what age you are, you need to be concerned about getting sufficient calcium in your diet. Your bones depend on it, and you depend on your bones for a lifetime. By supplying your bones with the calcium they need from the very beginning, you give your bones a chance to grow strong. By sustaining your body's need for calcium throughout your life, you help keep your bones from breaking down.

What if you've never paid much attention to calcium, and now, in your thirties, forties, or fifties, you've become concerned about osteoporosis? Is it too late to strengthen your bones by ensuring you get the calcium you need every day? Can the food you eat supply you with all the calcium you require, or do you need to take supplements? Are there other important risk factors you need to consider?

Let me begin by talking about calcium. Calcium is a mineral required for building teeth and bones, and necessary to maintain strong bones for life. In fact, 99 percent of the body's calcium is stored in your bones and teeth. When your bones don't get the calcium they need, they weaken.

It's never too early to be concerned about calcium, but it's especially important from adolescence on. A majority of women reach peak bone mass between 25 and 30; bone mass declines slowly at about 1 percent a year after then until menopause, at which point the rate of bone loss escalates to approximately 5 percent each year for about five years before slowing down again.

Calcium isn't the whole story, though. In order to absorb the calcium your bones and teeth require, you need sufficient Vitamin D. Most people get the Vitamin D they require (about 5–10 micrograms a day) from normal exposure to sunlight and by eating fortified foods. Most milk is fortified with Vitamin D, for instance. If you're not getting enough sunlight (during the winter, for example, or because you aren't outside enough during daylight hours) and/or you don't eat enough of these fortified calcium products, a daily multivitamin can provide you with your Recommended Daily Allowance (RDA) of 400 International Units (I.U.).

Other factors influence how your body uses the calcium you consume. Eating an excessive amount of animal protein can put you at risk. Why would a high-animal-protein diet affect your body's calcium use? When protein is digested, acids are produced by the body that draw calcium from your bones. So if you're consuming an excessive amount of protein, you're producing an abundance of these acids, and costing your body the calcium it so desperately requires.

Some medications also "steal" calcium from your body, including excessive thyroid hormone, steroids such as prednisone and cortisone (used to treat asthma and other disorders), lithium, anticonvulsants, and most chemotherapy drugs. If you're required to take any of these, it's important to ask your doctor how you can combat the loss of calcium during your treatment.

What Is Osteoporosis?

Osteoporosis is a "silent" disease that is characterized by a decrease in bone mass. When insufficient calcium is available, the bones begin to weaken and grow brittle. This condition, which produces so many devastating bone fractures in older women, develops over many years and usually progresses with few outward signs that something is wrong. Some older women actually grow shorter because they fracture vertebrae in their spines that collapse and/or compress. (This produces the condition most commonly known as "Dowager's Hump.") The most common injury associated with osteoporosis is a broken hip, a dangerous and sometimes life-threatening medical problem that requires complicated surgery and lengthy hospitalizations to heal. Other fractures of the spine may occur simply from routine activities like lifting, bending, or rising from a chair. These "compression" fractures can be extremely painful and therefore quite debilitating.

At a recent scientific summit, it was determined that more than 75 million people suffer from osteoporosis in Europe, Japan, and the United States, with more than 200 million projected worldwide. If current trends continue, those numbers are expected to double by the year 2020.

Are You at Risk for Osteoporosis?

If your mother or grandmother developed osteoporosis, you are more likely to be at risk. The good news is, the reverse is also true.

Some risk factors are out of your control. For example, women are at higher risk than men because they have less bone mass to begin with, and then enter menopause, which significantly accelerates the rate of bone mass decline. Caucasian women, and to a lesser extent Asian women, are at greater risk than African-American and Hispanic women, who tend to have denser bones. (But women of color should not ignore the issue of osteoporosis, wrongly believing they are not at risk.) Other studies suggest that the disease is more common in cold and temperate climates than in the tropics.

Not all the research emphasizes negative risk factors. There's also been encouraging data that indicates that women who've been pregnant and borne children derive benefit from the estrogen surges that occur during pregnancy and breastfeeding.

Here are some other important risk factors:

- Do you smoke?

- Do you drink caffeine or alcohol to excess?

- Do you lead a sedentary lifestyle?

- Do you get insufficient calcium from your diet?

- Do you get insufficient amounts of vitamins A and C or magnesium?

- Do you consume too much sodium (which increases the calcium you excrete)?

- Do you have or have you ever had an eating disorder like anorexia or bulimia?

- Do you follow a diet high in animal protein?

- Do you exercise to excess?

- Have you experienced amenorrhea (cessation of menstruation)?

- Do you take medications (like steroids) that deplete your body's calcium?

- Do you suffer from kidney failure?

- Are you being treated for parathyroid diseases, hyperthyroidism, rheumatoid arthritis, lupus, or a gastrointestinal disease where there is malabsorption, like colitis?

- Are you immobilized or restricted to a bed or wheelchair?

- Have you experienced early menopause (before age 45), or had early menopause for surgical reasons (i.e., removal of ovaries)?

- Are you going through menopause now?

If you answered "yes" to any of these questions, you may be at a high risk of developing osteoporosis. Predicting who will lose bone mass quickly is a complex effort, but you can protect yourself to a substantial extent by changing the way you live—and the way you nourish yourself. While you can't replenish what you may have lost, you *can* sustain the bone mass you have, and you *can* prevent further loss.

What Is the Recommended Daily Allowance (RDA) for Calcium?

Current recommendations suggest that women need to consume this much calcium every day:

Age	Milligrams of Calcium
0–10	800
11–24	1200
25 and older	800–1000
Pregnant	1200
Lactating (Breastfeeding)	1200
Postmenopausal	1500

(It should be noted that women who are either pregnant or breast-feeding require more calcium in their diet at these times. The reason is simple: during pregnancy, women must nourish not only their own bones but also the bones of the developing newborn. In order to grow sturdy, the baby's bones will take whatever calcium they need from Mom. Therefore, Mom has to make up that difference in her own diet, so that her own bones don't suffer. The need for additional calcium continues through the time she breastfeeds her infant, since calcium is further "lost" through her breast milk.)

Scientists continue to review and update their recommendations about consumption of important nutrients, but it often takes years before changes are adopted. A 1985 study by the National Institute of Health noted that most women in the United States consumed about half of the recommended allowance of calcium. The committee proposed that the RDA for adult women be increased to 1,200 to 1,500 milligrams per day.

Besides Milk Products, What Are Good Dietary Sources of Calcium?

Your best bets for getting calcium in your diet are a wide variety of dairy products, including milk, yogurt, and cheese. While the calcium in milk products is the same—calcium is calcium—low-fat milk products are a healthier choice. You may be surprised to learn, for example, that whole milk contains 3 percent fat, so that 2 percent milk, which is touted as a low-fat product, is only minimally lower in fat than whole milk.

Many other foods contain some calcium (see the chart in **A Guide to Calcium-Rich Foods, page 15**), but research has shown that the body tends to absorb calcium best from dairy foods or supplements when taken as directed. That's why the recipes in this cookbook emphasize the nutritional value and good taste provided most often by milk.

What If You're Lactose Intolerant?

If you have experienced difficulty consuming milk products, you may be lactose intolerant. This means that you lack a digestive enzyme required to absorb dairy products well. Many people with lactose intolerance can handle small amounts of dairy foods. Others use lactose-free milk products, or take a supplement that contains the required enzyme.

A number of over-the-counter products have been developed to help combat these problems, but none of them is perfect. Those that require you to add something to your dairy food may change the taste of that food; those you swallow may cause you to experience uncomfortable symptoms like gas and bloating.

Recent studies have shown that nearly 70 percent of African Americans and Asians are lactose intolerant. Dairy products are rarely consumed as part of an Asian diet, which emphasizes soy products as calcium sources. Tofu, which is available in most supermarkets and at Asian groceries, is made from soy milk and contains a mineral binder. Other good non-dairy sources of calcium include canned salmon and sardines. If you fall into this lactose-intolerant category and your diet is low in calcium, you need to be especially careful about consuming excessive caffeine or alcohol.

Can You Consume Too Much Calcium?

Sufficient calcium intake is vital for good bone health, but it's certainly possible to consume more than your body can use. And unlike some water-soluble vitamins (like C, for example) that simply "wash" out of your body if you take too much, too much calcium can lead to kidney stones and obstructions in your urinary tract. While it may be tempting to "self-dose" on calcium supplements, you should always check with your physician before exceeding the recommended daily allowances.

The Good News About Exercise and Osteoporosis

Regular exercise provides many different benefits, but only certain kinds of exercise help protect the health of your bones. These exercises, called "weight-bearing," require your body to work against gravity. Swimming, for example, is a great aerobic exercise, good for your muscles and your heart. But it won't help you build strong bones.

What are the best exercises for ensuring your bones get strong and stay that way? Here's a brief list:

- Walking

- Jogging

- Wearing wrist and ankle weights while walking or jogging

- Jumping rope

- Doing low-impact aerobics

- Exercising with weights to strengthen the upper body

Many women make time in their busy schedules for regular aerobic exercise, but they're less likely to get the kind of upper-body exercise that weight training delivers. Working out with hand weights helps to strengthen the back muscles responsible for keeping us upright; between the effects of gravity and the problems of osteoporosis, many women develop a stooped posture that is difficult to reverse. Women who take regular aerobic classes may experience bone loss in the upper body unless hand weights are added to their routines.

What if you're temporarily immobilized, or if you're restricted to a wheelchair or bed? Lack of activity causes your bones to break down and places you at serious risk. In fact, a day of bed rest is considered the equivalent of a year of living a sedentary lifestyle. If minor surgery or a bout of bronchitis keeps you confined to your bed, you may lose a substantial amount of conditioning in just a few days.

Check with your doctor, of course, but even if you're not ambulatory, you can usually perform many upper-body exercises. Focus on what you *can* do, whatever that is, and commit yourself to building bone strength.

If you can only walk a few blocks each day, do so. Don't become discouraged if it seems difficult at first. You'll soon be able to increase the amount of distance you can cover and the amount of weight you can lift. Remember this, though: Excessive exercise can also contribute to bone loss. Think of very slender dancers and runners who work out so intensely that their bodies no longer produce the hormones required for menstruation. When it comes to osteoporosis, you can be *too* thin!

A last note: Not all "weight-bearing" exercise is good for you *when you already suffer from osteoporosis.* Carrying heavy bags or even growing toddlers can put pressure on your spine and cause compression fractures. If your bones are already fragile, seemingly innocuous activities like hugging a loved one or twisting during a tennis match can fracture a rib.

The Role Estrogen Plays in Osteoporosis

Most people are unaware that estrogen is a major factor in keeping bones strong throughout a woman's life. Estrogen keeps calcium in the bones in three important ways.

First, it helps the intestines absorb as much calcium as possible from the foods you eat every day. Second, it discourages and slows the loss of calcium from your bones, which is a normal part of aging. Finally, it helps the kidneys to hold on to more calcium so that less is excreted into your urine.

The arrival of menopause in a woman's life signals the greatest danger to bone density and health. Why? Because menopause represents a major change in the body's estrogen level, as the ovaries cease working and estrogen levels drop. Estrogen is so critical to the prevention of bone loss that women who are at risk of developing osteoporosis usually require estrogen supplements once menopause occurs. Taking calcium helps, of course, but it is not nearly as effective as estrogen. Ongoing studies continue to demonstrate that women who do not take estrogen supplements suffer substantial bone loss.

Science writer and *New York Times* columnist Jane E. Brody recently reported on a study which demonstrated that estrogen replacement therapy (ERT) has actually helped to *increase* bone mass in the spine and hips in women who began taking the hormone in their sixties. If this success can be sustained, it would substantially reduce the risk of osteoporosis and resulting back and hip fractures. (An additional benefit of this therapy appears to be a reduction in cholesterol levels and a reduction in heart disease.)

It's important to consult a physician trained in evaluating your body's unique requirements for estrogen, as there is no standard dosage appropriate for every woman. And, not all women can take estrogen. A variety of factors need to be considered before estrogen is prescribed.

Some women have expressed concern about the long-term effects of estrogen supplements, noting that some studies have suggested an increased risk for developing some forms of cancer. But for women at risk of developing severe, debilitating osteoporosis—and this includes millions of women annually—estrogen supplements can literally be a lifesaver.

How Can You Find Out If You Have Osteoporosis?

The focus of this book, and of this introduction, is primarily on what all people can do to build the healthiest bones possible and sustain them for life. But if you consider yourself at serious risk of developing osteoporosis, or if your physician expresses concern about the health of your bones, then the strength of your bones may be tested by bone density scanning.

Low bone density means you're at risk for serious fracture, just as high blood pressure can indicate you're a candidate for a heart attack, or high blood pressure can warn that you're at risk for a stroke.

Bone density cannot be evaluated by ordinary X rays because early bone loss isn't sufficiently visible during that test. Instead, doctors use a technique called bone densitometry. The least expensive and most reliable test is called a DEXA (Dual-Energy X-ray Absorptiometry) scan, which measures bone density at a variety of sites around your body (usually your spine and hip, among others) and exposes you to only a very small amount of radiation. This remarkable test can spot even a 1 percent loss of bone density. During the test, the patient lies down for 10 to 15 minutes while a wand is passed over her body, providing a visible guide to the density of her bones. Subsequent scans can be compared with earlier scans to gauge if a chosen therapy has been successful.

Who should have the DEXA scans? In addition to post-menopausal women who can't or won't take estrogen, a recent *Newsweek* article suggested screening premenopausal women with bone-threatening conditions such as rheumatoid arthritis or hyperthyroidism, as well as "thin-boned" men over 65. (Remember, men get osteoporosis, too!) But anyone with one or more risk factors for osteoporosis (as previously listed) might consider being screened with a DEXA scan.

If DEXA scans are not available, some women are evaluated using a CAT scan. A CAT scan can only examine the spine, generally costs more, and requires more radiation to get the necessary results, so the DEXA is usually preferred. Osteoporosis is sometimes diagnosed during review of a chest X ray, but research shows that by the time bone loss shows up on a standard X ray, 30 percent of bone has already been lost.

Other evaluation techniques are being tested, including tests that can detect "biochemical markers" in urine and blood that indicate the rate at which bone is breaking down.

What Can Medicine Do If You Already Show Signs of Osteoporosis?

I've already discussed the value of lifestyle changes in preventing and postponing a loss of bone mass. Getting the calcium you need, doing regular weight-bearing exercise as well as building upper-body strength through weight training—those are steps that everyone can take to stay healthy.

But if you've got osteoporosis, your physician can offer you real help and hope through medication management. First, we'll work to correct any underlying medical conditions that are contributing to osteoporosis, such as an overactive thyroid or intestinal malabsorption (an inability to benefit from the nutrients you're consuming).

Second, if it's appropriate, your physician can decide to prescribe estrogen supplements to help you sustain bone density.

Finally, there are medicines to strengthen bones. There are now two types of drugs currently available for this purpose:

A) **Biphosphonates**, such as Fosamax, which block bone resorption (the breakdown of bones)

B) **Calcitonin**, which is a hormone that also blocks bone resorption and is now available for intranasal use (nose spray) as well as by injection

A third category of drugs under investigation for the future is the fluorides, which may help to increase new bone formation.

These drugs are prescribed for patients, both women and men, who can't take estrogen, or who have been taking estrogen but find that their bones are continuing to weaken and break down.

Your internist may decide to refer to you to a specialist—an endocrinologist or rheumatologist—who is specially trained to treat people with osteoporosis. Each individual needs to be carefully evaluated before treatment is prescribed.

What Can You Do to Prevent Osteoporosis?

Think of your best defense against osteoporosis as a five-part plan. Each one plays a vital role in assuring that your bones stay as strong as possible, for as long as possible!

1. Get the Calcium You Need—Daily.

The best way to get enough calcium is to eat plenty of easily absorbed calcium-rich foods and know you're getting the right amount of Vitamin D to help your body absorb them properly. Because it can sometimes be difficult to get all the calcium you need through diet alone, many people choose to take calcium supplements.

Before buying a bottle marked "Calcium" at your local drugstore or supermarket, be sure you understand that not all calcium provides the same value to your body. The tablet you swallow or chew (as in the case of antacids like Tums) isn't 100 percent calcium, but is combined with other ingredients. **Calcium carbonate,** which I suggest to my patients as the best form of oral supplement, is widely available in powder, pill, or antacid form. The average pill contains 500 milligrams of *elemental* calcium. Other forms of calcium include **calcium citrate** and **calcium gluconate.** But remember, it's the amount of *elemental* calcium contained in the supplement that counts!

You may see "chelated calcium" recommended as being more effectively absorbed, but recent research suggests it is no better than calcium without this additive. Other forms to avoid: bone meal or dolomite, which may contain toxic metals like lead.

You should take your calcium supplements **with meals,** as food increases the body's absorption of the mineral because digestion produces stomach acid. (This is one reason that calcium citrate can be a good supplement to take—it provides its own acidic environment.)

Don't take it at the same time as other medicines, however, as calcium can and often does interfere with how your body absorbs those prescribed drugs. Most physicians recommend splitting your calcium dose between morning and evening in order to help the body absorb as much as possible. This is particularly important if you are postmenopausal or going through menopause. Regrettably, as we age, we absorb just about everything less well, including nutrients from food. If you find yourself experiencing unpleasant symptoms like bloating, constipation, gas, or indigestion, spreading out your calcium supplements as well as your food should help to relieve the discomfort.

Lactose intolerance is often volume related, and as people age, they should probably reduce their intake of dairy foods to no more than 4 to 6 ounces at one time. You'll find that the recipes in this cookbook offer a "boost" of calcium but don't overload the digestive system.

2. Maintain a Healthy Body Weight.

Larger women actually have more bone mass than thinner women, which is one of the reasons that women who've suffered from anorexia or bulimia are very much at risk for developing osteoporosis. In fact, women with anorexia lose bone mass at a similar rate to menopausal women.

If you're overweight, you may find it difficult to begin and stick with a regular program of weight-bearing exercise. It may be encouraging to know that those extra pounds help build stronger bones. This is not to say that obesity is healthy—it's not. And, overweight women may develop osteoporosis too. But it can be better for your bones for you to be slightly overweight than underweight.

3. Exercise on a Regular Basis.

Find ways to increase the time you spend pounding the pavement, and take every opportunity to develop your upper-body strength. Park a few blocks away from the store when you shop, and remind yourself when you're carrying those bags of groceries that you're building bone (and muscle)! Regular exercise is good for your heart and your bones, so make it a part of your daily activities. You don't think about brushing your teeth; you just do it. The same should be true for fitting exercise into your healthy lifestyle.

4. Don't Smoke. (If You Do, Get the Help You Need.)

Recent research has shown that smoking isn't only bad for your heart and lungs, it's bad for your bones. In fact, smokers begin to lose bone density at a fast rate years before they reach menopause! Many women who smoke also experience menopause earlier than nonsmokers, and early onset of menopause is a distinct osteoporosis risk factor.

Smoking also influences how the body uses estrogen, and continued smoking during menopause and afterward can lessen the value of estrogen supplements.

5. Opt for Estrogen Replacement Once You Reach Menopause.

When menopause commences, your risk for osteoporosis is greatest. Estrogen replacement therapy has provided millions of women with the tools to combat this debilitating disease. If you have concerns about the effects of estrogen, discuss them with your doctor *before* you reach menopause, so that you'll be prepared to face the changes in your body and your bone health when the time comes.

Is It Ever Too Late to Prevent Osteoporosis?

Once bone mass is lost, it's difficult to rebuild it. But you can stop bone loss in its tracks by choosing a healthy lifestyle that includes getting sufficient calcium and exercise, quitting smoking, and beginning an estrogen replacement or other medication regimen. Even women in their seventies have seen positive results from making these changes.

Protecting the health of your bones is a lifetime job. I hope you'll share your concerns about osteoporosis with your family members, especially young women who are in their prime "bone-building" years or planning to start a family. And while most osteoporosis alerts are directed toward women, men also need to ensure long-term bone strength by meeting dietary requirements for calcium and making time for exercise.

The recipes in this cookbook provide lots of alternatives to drinking two glasses of skim milk each day to fulfill your RDA. By taking every opportunity you can to increase your family's calcium intake, you will begin to reap the many benefits of a healthy lifestyle.

—Brian L. Levy, M.D., F.A.C.E.

Now that you've heard from a doctor how adding calcium to your diet can keep your bones and teeth strong and healthy, I'm happy to share with you the following helpful information about ways to incorporate this wonderful bone-building mineral into your family menus.

—JoAnna

A Guide to Calcium-Rich Foods Courtesy of NATIONAL DAIRY COUNCIL®

Calcium's Value

Most people know that calcium is a mineral that helps build strong bones and teeth. But did you know that calcium also . . .

- helps your muscles contract and relax

- helps your heart beat

- helps your blood clot

- helps your nerves send messages

So what happens if you don't supply your body with enough calcium to perform these important functions? Your body takes the calcium it needs from your bones!

Your bones act as a kind of savings account for calcium. If your diet supplies enough calcium, your body deposits some in your bones. If your diet is low in calcium, your body makes a withdrawal from your bones.

Penalties for Not Keeping a Minimum Balance

A diet low in calcium has been linked to several health problems.

Osteoporosis—a crippling bone disease. Bones become so brittle that they break easily.
Bone loss in the jaw—this can lead to difficulty chewing, tooth loss, and poor-fitting dentures.
Hypertension—high blood pressure can lead to strokes and heart attacks in some people.

How to Make a Deposit

If you're like most people, you may not be getting all the calcium you need. To provide your body with enough calcium, you need to make daily "deposits."

*Suggested Daily Calcium Deposit

Age/Stage	Calcium
1–10	800 mg
11–24	1200 mg
25 and older	800 mg
Pregnant and Lactating	1200 mg

It's easy to make this deposit. And you have dozens of calcium-rich options to choose from. Look over the lists of foods on the following pages. Check the foods you'd be willing to add to your diet.

Over Fifty Calcium-Rich Ways to Make a Deposit

Milk Group

Most of the calcium in the food supply comes from foods in the Milk Group. It's very difficult to get all the calcium you need without eating Milk Group foods.

*The National Institutes of Health Expert Panel, 1994, recommends that adults consume 1,000–1,500 mg of calcium per day.

While the Milk Group supplies 75 percent of the calcium in our food supply, it only supplies 12 percent of the fat. If you're watching your fat intake, look for foods with an asterisk (*).

Other Assets: also a major source of riboflavin, protein, and vitamin D.

Your Calcium Options	Milligrams of Calcium
Yogurt, plain, nonfat (1 cup)*	452
Yogurt, plain, lowfat (1 cup)*	415
Cheese, Swiss (1½ oz)	408
Milkshake, chocolate (10 fl oz)	319
Yogurt, fruit-flavored, lowfat (1 cup)*	314
Cheese, cheddar (1½ oz)	306
Milk, skim (1 cup)*	302
Milk, 1% lowfat (1 cup)*	300
Milk, 2% lowfat (1 cup)*	297
Milk, whole (1 cup)*	291
Buttermilk (1 cup)*	285
Milk, chocolate, 2% lowfat (1 cup)	284
Milk, chocolate, whole (1 cup)	280
Cheese, mozzarella, part skim (1½ oz)	275
Cheese, American (1½ oz)	261
Pudding, cooked (½ cup)	152
Ice cream, soft serve (½ cup)	113
Yogurt, frozen, vanilla (½ cup)*	103
Ice cream, hardened, 16% fat (½ cup)	87
Ice cream, hardened, 10% fat (½ cup)	85
Cheese, cottage, 2% lowfat (½ cup)*	77

* Lower in fat (5 grams of fat or less per serving)

Meat, Vegetable, and Grain Groups

The Meat Group supplies about 9 percent of the calcium in the food supply. The Vegetable and Grain Groups supply even less.

A few foods in each of these groups contain calcium. For example, canned salmon and sardines supply calcium, when you eat the bones. Some leafy green vegetables contain calcium, too. So do some grain products.

Meat Group

Other Assets: also a major source of protein, niacin, iron, and thiamin.

Your Calcium Options	Milligrams of Calcium
Tofu, with calcium sulfate (½ cup)	434
Sardines, canned, with bones (3 oz)	321
Salmon, canned, with bones (3 oz)	203
Tofu, without calcium sulfate (½ cup)	130
Perch, baked (3 oz)*	117
Almonds (⅓ cup)	114

Vegetable Group

Other Assets: also a major source of vitamin A, vitamin C, and fiber.

Spinach, fresh, cooked (½ cup)†	122
Turnip greens, fresh, cooked (½ cup)*	99
Okra, frozen, cooked (½ cup)*	88
Beet greens, fresh, cooked (½ cup)*	82

Grain Group

Other Assets: also a major source of carbohydrate, thiamin, iron, niacin, and fiber.

Waffle, homemade (7" waffle)	191
Biscuit, from mix (1 biscuit)	105

* Lower in fat (5 grams or less per serving)
† The calcium in spinach is very poorly absorbed.

Combination Foods/Fast Foods, and the "Others" Category

Combination Foods are made with foods from more than one food group. It's the foods from the Milk Group that make these Combination Foods good sources of calcium.

Foods in the "Others" category are low in most nutrients and are usually high in calories.

Combination Foods/Fast Foods
Other Assets: good source of many other nutrients.

Your Calcium Options	Milligrams of Calcium
Lasagna (2½" × 2¼")	460
Macaroni and cheese, homemade (1 cup)	362
Enchilada, cheese (1 enchilada)	324
Taco Bell's® Chili Cheese Burrito	300
Chef's salad, without dressing (1½ cups)	235
Quiche (1/8 pie)	211
Pizza, meat and vegetable, thin crust (¼ of 12")	201
McDonald's Big Mac® (1 sandwich)	200
Cheeseburger, regular (1 sandwich)	182
Tomato soup, with milk (1 cup)	159
McDonald's Egg McMuffin® (1 sandwich)	150
Burger King's Croissan'wich® (1 sandwich)	150
Spaghetti with meatballs (1 cup)	124
Wendy's® broccoli and cheese potato (1)	100
McDonald's Filet-o-Fish® (1 sandwich)	100
Submarine sandwich (3"–4" sub)	95

"Others" Category
Low in most nutrients

Pie, chocolate cream (1/8 of 9" pie)	115

Calcium-Fortified Foods
Calcium is sometimes added to orange juice, bread, soft drinks, cereal, milk, yogurt, and other foods. The amount of calcium in these foods is listed on the label. Keep in mind that calcium-fortified foods can increase your calcium intake. However, they do not provide the body with other nutrients supplied by dairy foods.

Calcium Supplements

There is a wide variety of calcium supplements on the market. The amount of calcium in these supplements is listed on the label. There are no benefits for taking more than your RDA for calcium. In fact, high doses of calcium may interfere with the absorption of other nutrients, like iron. The National Academy of Science recommends that individuals not consume more than 1,000 milligrams of calcium per day from supplements.

Talk to your physician or dietitian before taking a calcium supplement.

The American Medical Association and the American Dietetic Association recommend that you get your calcium from food, rather than from pills.

Pills are not a substitute for a nutritionally adequate diet.

Bankruptcy Prevention

Getting enough calcium is not that difficult.* Here are some easy ways to add more calcium to your diet.

- For breakfast, prepare your hot cereal or cocoa with milk, instead of water.

- For lunch, fill a melon with cottage cheese or frozen yogurt.

- For dinner, toss grated cheese on your salad or baked potato.

- For a snack, enjoy yogurt. Add fruit, nuts, or granola if you like.

Of course, you cannot live on calcium alone. Your body needs more than forty nutrients to function properly. But no one food—or food group—supplies them all. That's why you need a variety of foods from the Five Food Groups every day.

*If you're lactose intolerant, look for the "DAIRY COUNCIL's Getting Along with Milk" brochure.

Recommended Daily Servings

Milk Group*

• children (1–10)	3
• teenagers and young adults (11–24)	4
• adults (25+)	2
• pregnant and breastfeeding women	4

Meat Group	2–3
Vegetable Group	3–5
Fruit Group	2–4
Grain Group	6–11

For healthy benefits today and in years to come—keep making regular, daily calcium deposits.

*To meet the calcium recommendations issued by a 1994 National Institutes of Health Expert Panel, children should consume 3–4 servings of milk; teens and young adults should consume 4–5 servings; adults should consume 3–5 servings; and pregnant and breastfeeding women should consume 4–5 servings.

"A Guide to Calcium-Rich Foods" was originally published as "The All-American Guide to Calcium-Rich Foods."

How I Learned
to Help Myself

JoAnna M. Lund and
Healthy Exchanges

For twenty-eight years I was the diet queen of DeWitt, Iowa. I tried every diet I ever heard of, every one I could afford, and every one that found its way to my small town in eastern Iowa. I was willing to try anything that promised to "melt off the pounds," determined to deprive my body in every possible way in order to become thin at last.

I sent away for expensive "miracle" diet pills. I starved myself on the Cambridge Diet and the Bahama Diet. I gobbled Ayds diet candies, took thyroid pills, fiber pills, prescription and over-the-counter diet pills. I went to endless weight-loss support group meetings—but I somehow managed to turn healthy programs such as Overeaters Anonymous, Weight Watchers, and TOPS into unhealthy diets . . . diets I could never follow for more than a few months.

I was determined to discover something that worked long-term, but each new failure increased my desperation that I'd never find it.

I ate strange concoctions and rubbed on even stranger potions. I tried liquid diets like Slimfast and Metrecal. I agreed to be hypnotized. I tried reflexology and even had an acupuncture device stuck in my ear!

Does my story sound a lot like yours? I'm not surprised. No wonder the weight-loss business is a billion-dollar industry!

Every new thing I tried seemed to work—at least at first. And losing that first five or ten pounds would get me so excited, I'd believe that this new miracle diet would, finally, get my weight off for keeps.

Inevitably, though, the initial excitement wore off. The diet's routine and boredom set in, and I quit. I shoved the pills to the back of the medicine chest; pushed the cans of powdered shake mix to the rear of the kitchen cabinets; slid all the program materials out of sight under my bed; and once more I felt like a failure.

Like most dieters, I quickly gained back the weight I'd lost each time, along with a few extra "souvenir" pounds that seemed always to settle around my hips. I'd done the diet-lose-weight-gain-it-all-back "yo-yo" on the average of once a year. It's no exaggeration to say that over the years I've lost 1,000 pounds—and gained back 1,150 pounds.

Finally, at the age of forty-six I weighed more than I'd ever imagined possible. I'd stopped believing that any diet could work for me. I drowned my sorrows in sacks of cake donuts and wondered if I'd live long enough to watch my grandchildren grow up.

Something had to change.

I had to change.

Finally, I did.

I'm just over fifty now—and I'm 130 pounds less than my all-time high of close to 300 pounds. I've kept the weight off for more than six years. I'd like to lose another ten pounds, but I'm not obsessed about it. If it takes me two or three years to accomplish it, that's okay.

What I *do* care about is never saying hello again to any of those unwanted pounds I said good-bye to!

How did I jump off the roller coaster I was on? For one thing, I finally stopped looking to food to solve my emotional problems. But what really shook me up—and got me started on the path that changed my life—was Operation Desert Storm in early 1991. I sent three children off to the Persian Gulf war—my son-in-law Matt, a medic in Special Forces; my daughter Becky, a full-time college student and member of a medical unit in the Army Reserve; and my son James, a member of the Inactive Army Reserve reactivated as a chemicals expert.

Somehow, knowing that my children were putting their lives

on the line got me thinking about my own mortality—and I knew in my heart the last thing they needed while they were overseas was to get a letter from home saying that their mother was ill because of a food-related problem.

The day I drove the third child to the airport to leave for Saudi Arabia, something happened to me that would change my life for the better—and forever. I stopped praying my constant prayer as a professional dieter, which was simply "Please, God, let me lose ten pounds by Friday." Instead, I began praying, "God, please help me not to be a burden to my kids and my family."

I quit praying for what I wanted and started praying for what I needed—and in the process my prayers were answered. I couldn't keep the kids safe—that was out of my hands—but I could try to get healthier to better handle the stress of it. It was the least I could do on the homefront.

That quiet prayer was the beginning of the new JoAnna Lund. My initial goal was not to lose weight or create healthy recipes. I only wanted to become healthier for my kids, my husband, and myself.

Each of my children returned safely from the Persian Gulf war. But something didn't come back—the 130 extra pounds I'd been lugging around for far too long. I'd finally accepted the truth after all those agonizing years of suffering through on-again, off-again dieting.

There are no "magic" cures in life.

No "magic" potion, pill, or diet will make unwanted pounds disappear.

I found something better than magic, if you can believe it. When I turned my weight and health dilemma over to God for guidance, a new JoAnna Lund and Healthy Exchanges were born.

I discovered a new way to live my life—and uncovered an unexpected talent for creating easy "common folk" healthy recipes, and sharing my commonsense approach to healthy living. I learned that I could motivate others to change their lives and adopt a positive outlook. I began publishing cookbooks and a monthly food newsletter, and speaking to groups all over the country.

I like to say, "When life handed me a lemon, not only did I make healthy, tasty lemonade, I wrote the recipe down!"

What I finally found was not a quick fix or a short-term diet, but a great way to live well for a lifetime.

I want to share it with you.

What Healthy

Exchanges Means

I s it really possible to enjoy favorite foods once the excess sugars and fats are removed? Yes, and I can prove it to you—just as I proved to myself and my family that I could create healthy recipes that tasted like "real food."

When I came up with the concept for Healthy Exchanges, I had three ideas in mind:

1. I wanted to "**exchange**" old, unhealthy habits for new, healthy ones in food, exercise, and mental attitude.

2. I chose to "**exchange**" ingredients within each recipe to eliminate as much fat and sugar as possible, while retaining the original flavor, appearance, and aroma.

3. I calculated all recipes using the **exchange** system of measuring daily food intake as established by the American Dietetic Association, the American Diabetic Association, and many national weight-loss organizations.

The Strong Bones Healthy Exchanges Cookbook can help you accomplish your healthy living goals, whatever they are. Maybe you're concerned about developing osteoporosis. Maybe you want to eat healthy. Maybe you need to stabilize your blood sugar better. Maybe you need to lower your cholesterol. Or maybe you need to lose a few pounds—or a few more than that. But you need to find delicious, easy-to-make recipes the rest of the family will eat without objection. The dieter or health-conscious person may be willing to accept uninspired, low-fat, low-sugar "diet" recipes—but what can you do when your kids want pizza and your husband insists on hot fudge sundaes and pie?

Most diet recipes deserve their bad reputations. Too many of them sacrifice taste and appeal for lower calories and less sodium or sweetener. That's usually why nobody likes to eat them—and why even if we stick to this kind of eating for a while, we eventually jump off the wagon!

This book will change that—and change it for good. Ever since I began creating recipes, I aimed for what I call "common folk healthy"—meaning that the recipes I create have to fit into an already busy lifestyle, they have to consider our likes and dislikes as a family, and they have to appeal to everyone sitting at my table.

I do not shop at health food stores, and I don't like preparing complicated dishes that keep me in the kitchen for hours. I got tired of being left with a pile of dirty dishes to show for all my efforts once the family had bolted from the table after a mere ten minutes. People may laugh when I say, If it takes longer to fix it than it does to eat it, forget it—but you'll be amazed to discover that you can eat well *and* healthily without spending hours cooking, chopping, and cleaning up!

My goal is this: the foods that I prepare have to look, taste, smell, and feel like the foods we have always enjoyed. Because I figured out how to "exchange" the excess fats and sugars within each recipe with readily available healthy ingredients, you'll be able to enjoy "real-food" dishes like pizza, mashed potatoes, and cherry cheesecake. (See the Recipes section for details!) Besides reducing the overall fat and sugar content, I boosted the calcium and lowered the sodium content, too.

Using my recipes can help you build strong bones, lose weight, lower and control your cholesterol and/or blood sugar levels, and still allow you to enjoy delicious foods that deliver good nutrition and great "eye" appeal!

I didn't set out to impress anyone with fancy gourmet recipes that require unusual ingredients. Most of the mail I receive from people who've tried my recipes tells me that I made the right decision. They wanted (as my family and I did) to discover quick, tummy-pleasing, "common-folk" dishes that were healthy, too—and in my Healthy Exchanges recipes they found exactly what they were looking for.

My sister Jeanie is another good example of the Healthy Exchanges recipe user. She could probably best be described as a "non-cooking" cook. If a recipe isn't easy, she doesn't want to be bothered, because she has more important things to do with her time. But she gladly takes the few minutes required to stir up her favorite Healthy Exchanges recipes!

The Four "Musts"

Before I included any recipe in this book, it had to meet my Four "Musts." I knew that if I didn't address these "Musts" from the very beginning, I would be just dieting, and only days or minutes away from falling off the "health wagon," as I call it—and into a binge where I would be keeping company again with cake donuts and hot fudge sundaes (my personal favorites . . . and downfalls!).

By eliminating unnecessary preparation and excess "bad" foods, I've made it much easier to live and eat healthy. Each of my recipes must be:

1. **Healthy**. Every recipe is low in sugar and fat and within a reasonable range for sodium. And because it is all of these, it is also low in calories and cholesterol. The prepared food can be eaten with confidence by diabetics and heart patients as well as anyone interested in losing weight or maintaining a weight loss.

2. **Easy to make**. Most people don't have the time or the inclination for complicated recipes. They just want to prepare something quickly and get it on the table as fast as possible. But they still want their families to be smiling when they leave the table. Simply put, they want dishes that are quick and easy to prepare—but that don't look like it!

3. **As tasty and good as they look**. If the foods we eat don't look, taste, smell, and feel like the family favorites we are

used to eating, we won't be willing to eat them week in and week out.

We are all creatures of habit. If we grew up eating fried chicken with mashed potatoes every Sunday at noon, we may try poached chicken with plain potatoes once—but we'll go running back to our favorite greasy chicken as soon as we can get our hands on some. However, if we could enjoy healthy yet delicious oven-fried chicken with healthy yet satisfying mashed potatoes topped with rich, thick, healthy gravy, we would probably agree to give up the deep-fried chicken dinner. Rather than demanding we give up a beloved comfort food but instead serving a healthy version with all the great flavor of the original dish we remember, we can feel satisfied—and willing to make this "Healthy Exchange" for the rest of our lives.

Here's another example. "Good enough" will never be good enough for me again. My food has to be garnished in easy yet attractive ways so I feed my eyes as well as my stomach. That's why I sprinkle two tablespoons of chopped almonds on top of my **Strawberry Chocolate Cream Pie.** This dish serves eight, which works out to about ⅔ teaspoon of nuts per person. But we definitely taste that crunch as we enjoy that piece of dessert. And those delectable bites of pecan help keep us from gobbling handfuls of nuts while watching TV. It's when we think we can't have something that we *dwell* on it constantly until we do get it. Moderation is the cornerstone of Healthy Exchanges recipes. My recipes include tiny amounts of mini chocolate chips, coconut, miniature marshmallows, nuts, and other foods I like to call "real-people" foods—not so-called "diet" food. This is just one of the distinctions that set my cookbook apart from others that emphasize low-fat and/or low-sugar foods.

4. **Made from ingredients found in DeWitt, Iowa.** If the ingredients can be found in a small town with a population of 4,500 in the middle of an Iowa cornfield, then anyone ought to be able to find them, no matter where they shop! To make these recipes, you do not have to drive to a specialty store in a large city, or visit a health food store in search of special ingredients.

Making It Quick, Easy, and More . . .

Once a recipe passes the Four Musts, I have a few additional requirements. One includes using the entire can if the container has to be opened with a can opener. The reason for this is simple: Only a few people are perfect Suzy and Homer Homemakers; most people are like me. If a recipe called for ⅓ cup tomato sauce, I would put the rest in the refrigerator with good intentions to use it up within a day or two—and six weeks later would probably have to toss it out because it was covered with moldy globs of green stuff. When I create a recipe, I devise a way to use the entire can and still have the dish taste delicious (and add up right when it comes to the mathematics of the exchanges for each serving).

Another test is what I call **"Will It Play in Anytown, USA?"** I ask myself if the retired person living on a meager pension or the young bride living on a modest paycheck can afford to prepare the recipe in both equipment and ingredients. I ask myself if the person with arthritic hands or the mother with two small children underfoot can physically manage to make the recipe. If the answers are yes, that recipe will be included in my cookbooks or newsletters.

It's important to me that my recipes are simple to make and understand.

Here are some of the ways I do this:

- I provide both the weight and the closest cup or spoon measurement for each ingredient. Why? Well, everyone who follows the major diet plans or uses the American Diabetic Association program is required to use a kitchen scale every day. But I know that in the real world this just won't happen. People say they're too busy to pull the scale out of the cupboard. Or they'd rather not use the measuring bowl and have to wash it. So my recipe says that 3 ounces shredded reduced-fat cheese is equivalent to ¾ cup. I've measured this so often to make sure it's right—why should you have to as well?

- I measure the final product and give the closest cup measurement per serving for anything that can't be cut into exact portions. For example, in a soup recipe, the serving size might be 1½ cups.

- I suggest baking all casseroles in an 8-by-8-inch pan for ease of portion control. If the dish serves 4, just cut it down the center, turn the dish and cut it down the center again. Each cut square yields one serving. Many cookbooks call for a 1½-quart casserole, but I find that most people tend to scoop out too much or too little while the rest of the casserole collapses into the center. There's no guesswork my way—just an easy and accurate way to manage portion size.

Measuring Success
One Exchange at a Time

Healthy Exchanges recipes are unique for another reason: They provide nutritional information calculated in three ways.

1. **Weight Loss Choices™/Exchanges.** The recipes can be used by anyone attending the national weight loss programs or support groups that count daily food intake by exchanges or selections instead of calories. (For a complete explanation of Healthy Exchanges Weight Loss Choices/ Exchanges, please refer to *The Healthy Exchanges Cookbook.*)

2. **Calories, Fat Grams, and Fiber.** I list the fat grams right next to the calories. In traditional recipes, fat grams usually appear in the middle of the nutrient information. Now, if a person chooses to count fat grams or compare the percentage of fat calories to total calories, the information is quickly found. I also provide fiber grams so you can monitor your fiber intake if you choose to. My calculations include protein, carbohydrate, sodium, and calcium, too.

3. **Diabetic Exchanges.** A Registered Dietitian has calculated the Diabetic Exchanges to conform to the guidelines established by the American Diabetic Association. This makes this book a wonderful resource for diabetics, for those who

need to cook for diabetic family members, and for professionals who educate diabetics in how to eat healthfully and control their condition.

I've provided these three different kinds of nutritional measurements to make it easier for anyone and everyone to use my recipes as part of their commitment to a healthy lifestyle. I will do just about anything to make cooking healthier and easier for people—except do their dishes!

Is Healthy Exchanges for You?

Do you want to lose weight— and keep it off for good?

I am living proof that you can lose weight while eating dishes prepared from Healthy Exchanges recipes. I went from a much-too-tight size 28 to a healthy size 12 to 14 when I began eating my own creations. I always make sure the bases are covered when it comes to meeting nutritional needs, and *then* I throw in tiny amounts of delicious treats—foods like mini chocolate chips that are usually considered "no-no's" in traditional diet recipes. Yes, we can get by without these little touches, but do we want to? Should we have to?

When I was a "professional" dieter, a typical lunch would consist of a couple of slices of diet bread, tuna packed in water, diet mayonnaise, lettuce, a glass of skim milk, and a small apple. Good healthy choices, all of them. But why was I always reaching for a candy bar or cake donuts just an hour later? Because I had fed only my stomach, and not my heart, my eyes, or my soul. It was only when I began enjoying a healthy piece of a real dessert every day that I began to lose weight and keep it off permanently. When I finally said NO to diets, I said YES to lasting weight loss!

The cardinal sin most people commit when going on a diet is preparing two different meals—skimpy get-the-weight-off-as-fastas-possible food for the dieter, and real food for the rest of the family. My recipes can be eaten with confidence by anyone who wants to lose weight, but they are so tasty the entire family will gladly eat the same food. (I call it passing the "Cliff Lund Taste Test!") When you cook with Healthy Exchanges, the burden of preparing two separate meals, not to mention the temptation of "sampling" the good stuff to be sure "it tastes right," has been eliminated. Now instead of short-term dieting and weight loss that never seems to last, you can take a step forward to healthy living for a lifetime!

Are you a heart disease patient? Or do you just need to lower or limit cholesterol?

My Healthy Exchanges recipes are all low in fat, just right for you if you're recovering from a heart attack or if improving your cholesterol count is your goal. I have a recipe for **Pam's Lasagna** which calls for real ingredients (*not* dietetic) and contains only 186 calories with 6 grams of fat per serving. It tastes just as good as a traditional recipe which delivers three times the amount of fat and almost twice the calories. And I don't skimp on the serving size! My approach provides a real savings in calories and fat without sacrificing flavor. While my own blood pressure and cholesterol are both within normal range, I know how important it is to eat a diet low in fat. My parents both died of heart complications, so I've always been concerned about fat intake. This program provides dozens of delicious low-fat recipes to help you reach the best health possible.

Are you diabetic? Do you have hypoglycemia?

My recipes are always low in sugar. A typical serving of pumpkin pie prepared in the traditional way provides approximately 300 calories with 18 grams of fat per serving. My recipe for **Magical Pumpkin Pie** is *192 calories per serving* with only 8 grams of fat and *no* added sugar! Just because I banished excess fats and sugars doesn't mean I've sacrificed any flavor. Besides, my pie can be cut into 8 pieces instead of the usual 12 servings per pie suggested in many diet recipes. (Did you know that one-sixth of a pie used to be the standard-size serving? Now, too many recipes expect you to settle for one-twelfth of a pie! But that skinny sliver leaves you hungry, and going back for another piece. You're back to one-sixth of the pie again—and that might leave your tummy groaning and over-stuffed. But one-eighth of a pie is a satisfying, refreshing serving that will leave your palate happy!)

With Healthy Exchanges, diabetic and hypoglycemic patients can find the "real world" food, especially desserts, they've been missing. You see, while my own blood sugar is normal, both my parents and two uncles developed adult-onset diabetes. So creating delicious low-sugar recipes is a priority for me.

Are you simply interested in preventing medical problems like osteoporosis from developing? Are you ready to make a commitment to eat healthier?

Maybe you're lucky enough to have no immediate medical concerns, but you would like to keep from developing them in the future. Perhaps you've seen all too clearly what the burdens of compromised health caused by a poor diet can be. You don't want to inherit the same "complications" you may have observed in parents and older family members, and you're determined to do your best to avoid future health problems by taking the necessary steps now. Healthy Exchanges will help you ensure lifelong good health.

Are you interested in quick and easy recipes the entire family will love?

Maybe you don't really care if a recipe is healthy or not. Maybe you have other priorities right now in your life, and you don't want to spend any more time than necessary in the kitchen. But because you and your family have to eat, you want a collection of good-tasting dishes that can be "thrown together" without much fuss. I will give you recipes for healthy, delicious food that often can be prepared in five minutes or less. (Not including unattended cooking time, but that time is freed for other projects!)

Did you answer "yes" to any of those questions? If you did, then **Healthy Exchanges** will give you just what you need—and deliciously, too!

JoAnna's Ten Commandments of Successful Cooking

A very important part of any journey is knowing where you are going and the best way to get there. If you plan and prepare before you start to cook, you should reach mealtime with foods to write home about!

1. **Read the entire recipe from start to finish** and be sure you understand the process involved. Check that you have all the equipment you will need *before* you begin.

2. **Check the ingredient list** and be sure you have *everything* and in the amounts required. Keep cooking sprays handy—while they're not listed as ingredients, I use them all the time (just a quick squirt!).

3. **Set out *all* the ingredients and equipment needed** to prepare the recipe on the counter near you *before* you start. Remember that old saying, *A stitch in time saves nine*? It applies in the kitchen, too.

4. **Do as much advance preparation as possible** before actually cooking. Chop, cut, grate, or whatever is needed to prepare the ingredients and have them ready before you

start to mix. Turn the oven on at least 10 minutes before putting food in to bake, to allow the oven to preheat to the proper temperature.

5. **Use a kitchen timer** to tell you when the cooking or baking time is up. Because stove temperatures vary slightly by manufacturer, you may want to set your timer for 5 minutes less than the suggested time just to prevent overcooking. Check the progress of your dish at that time, then decide if you need the additional minutes or not.

6. **Measure carefully.** Use glass measures for liquids and metal or plastic cups for dry ingredients. My recipes are based on standard measurements. Unless I tell you it's a scant or full cup, measure the cup level.

7. **For best results, follow the recipe instructions exactly.** Feel free to substitute ingredients that *don't tamper* with the basic chemistry of the recipe, but be sure to leave key ingredients alone. For example, you could substitute sugar-free instant chocolate pudding for sugar-free butterscotch instant pudding, but if you used a six-serving package when a four-serving package was listed in the ingredients, or you used instant when cook-and-serve is required, you won't get the right result.

8. **Clean up as you go.** It is much easier to wash a few items at a time than to face a whole counter of dirty dishes later. The same is true for spills on the counter or floor.

9. **Be careful about doubling or halving a recipe.** Though many recipes can be altered successfully to serve more or fewer people, *many cannot.* This is especially true when it comes to spices and liquids. If you try to double a recipe that calls for 1 teaspoon pumpkin-pie spice, for example, and you double the spice, you may end up with a too-spicy taste. I usually suggest increasing spices or liquid by 1½ times when doubling a recipe. If it tastes a little bland to you, you can increase the spice to 1¾ times the original amount the next time you prepare the dish. Remember:

You can always add more, but you can't take it out after it's stirred in.

The same is true with liquid ingredients. If you wanted to **triple** a recipe like my **Broccoli Noodle Casserole** because you were planning to serve a crowd, you might think you should use three times as much of every ingredient. Don't, or you could end up with Broccoli Noodle Soup! The original recipe calls for 12 ounces of Carnation Evaporated Skim Milk, so I'd suggest using 24 ounces when you **triple** the recipe (or 18 ounces if you **double** it). You'll still have a good-tasting dish that won't run all over the plate.

10. **Write your reactions next to each recipe once you've served it.**

Yes, that's right, I'm giving you permission to write in this book. It's yours, after all. Ask yourself: Did everyone like it? Did you have to add another half teaspoon of chili seasoning to please your family, who like to live on the spicier side of the street? You may even want to rate the recipe on a scale of 1★ to 4★, depending on what you thought of it. (Four stars would be the top rating—and I hope you'll feel that way about many of my recipes.) Jotting down your comments while they are fresh in your mind will help you personalize the recipe to your own taste the next time you prepare it.

Healthy

Exchanges®

Weight Loss

Choices™/

Exchanges

If you've ever been on one of the national weight-loss programs like Weight Watchers or Diet Center, you've already been introduced to the concept of measured portions of different food groups that make up your daily food plan. If you are not familiar with such a system of weight loss choices or exchanges, here's a brief explanation. (If you want or need more detailed information, you can write to the American Dietetic Association or the American Diabetes Association for comprehensive explanations.)

The idea of food exchanges is to divide foods into basic food groups. The foods in each group are measured in servings that have comparable values. These groups include Proteins/Meats, Breads/Starches, Vegetables, Fats, Fruits, Skim Milk, Free Foods, and Optional Calories.

Each choice or exchange included in a particular group has about the same number of calories and a similar carbohydrate, protein, and fat content as the other foods in that group. Because any food on a particular list can be "exchanged" for any other food in that group, it makes sense to call the food groups *exchanges* or *choices*.

I like to think we are also "exchanging" bad habits and food choices for good ones!

By using Weight Loss Choices or exchanges you can choose from a variety of foods without having to calculate the nutrient value of each one. This makes it easier to include a wide variety of foods in your daily menus and gives you the opportunity to tailor your choices to your unique appetite.

If you want to lose weight, you should consult your physician or other weight-control expert regarding the number of servings that would be best for you from each food group. Since men generally require more calories than women, and since the requirements for growing children and teenagers differ from adults, the right number of exchanges for any one person is a personal decision.

I have included a suggested plan of weight loss choices in the pages following the exchange lists. It's a program I used to lose 130 pounds, and it's the one I still follow today.

(If you are a diabetic or have been diagnosed with heart problems, it is best to meet with your physician before using this or any other food program or recipe collection.)

Food Group Weight Loss Choices/Exchanges

Not all food group exchanges are alike. The ones that follow are for anyone who's interested in weight loss or maintenance. If you are a diabetic, you should check with your health-care provider or dietitian to get the information you need to help you plan your diet. Diabetic exchanges are calculated by the American Diabetic Association, and information about them is provided in *The Diabetic's Healthy Exchanges Cookbook* (Perigee Books).

Every Healthy Exchanges recipe provides calculations in three ways:

- Weight Loss Choices/Exchanges

- Calories, Fat, Protein, Carbohydrates, and Fiber Grams, and Sodium in milligrams

- Diabetic Exchanges calculated for me by a Registered Dietitian

Healthy Exchanges recipes can help you eat well and recover your health, whatever your health concerns may be. Please take a few minutes to review the exchange lists and the suggestions that follow on how to count them. You have lots of great eating in store for you!

Proteins

Meat, poultry, seafood, eggs, cheese, and legumes.
One exchange of Protein is approximately 60 calories. Examples of one Protein choice or exchange:

1 ounce cooked weight of lean meat, poultry, or seafood
2 ounces white fish
1½ ounces 97% fat-free ham
1 egg (limit to no more than 4 per week)
¼ cup egg substitute
3 egg whites
¾ ounce reduced-fat cheese
½ cup fat-free cottage cheese
2 ounces cooked or ¾ ounces uncooked dry beans
1 tablespoon peanut butter (also count 1 fat exchange)

Breads

Breads, crackers, cereals, grains, and starchy vegetables. One exchange of Bread is approximately 80 calories. Examples of 1 Bread choice or exchange:

1 slice bread or 2 slices reduced-calorie bread (40 calories or less)
1 roll, any type (1 ounce)

½ cup cooked pasta or ¾ ounce uncooked (scant ½ cup)
½ cup cooked rice or 1 ounce uncooked (⅓ cup)
3 tablespoons flour
¾ ounce cold cereal
½ cup cooked hot cereal or ¾ ounce uncooked (2 tablespoons)
½ cup corn (kernels or cream-style) or peas
4 ounces white potato, cooked, or 5 ounces uncooked
3 ounces sweet potato, cooked, or 4 ounces uncooked
3 cups air-popped popcorn
7 fat-free crackers (¾ ounce)
3 (2½-inch squares) graham crackers
2 (¾-ounce) rice cakes or 6 mini
1 tortilla, any type (6-inch diameter)

Fruits

All fruits and fruit juices. One exchange of Fruit is approximately 60 calories. Examples of 1 Fruit choice or exchange:

1 small apple or ½ cup slices
1 small orange
½ medium banana
¾ cup berries (except strawberries and cranberries)
1 cup strawberries or cranberries
½ cup canned fruit, packed in fruit juice or rinsed well
2 tablespoons raisins
1 tablespoon spreadable fruit spread
½ cup apple juice (4 fluid ounces)
½ cup orange juice (4 fluid ounces)
½ cup applesauce

Skim Milk

Milk, buttermilk, and yogurt. One exchange of Skim Milk is approximately 90 calories. Examples of 1 Skim Milk choice or exchange:

1 cup skim milk
½ cup evaporated skim milk
1 cup low-fat buttermilk
¾ cup plain fat-free yogurt
⅓ cup nonfat dry milk powder

Vegetables

All fresh, canned, or frozen vegetables other than the starchy vegetables. One exchange of Vegetable is approximately 30 calories. Examples of 1 Vegetable choice or exchange:

½ cup vegetable
¼ cup tomato sauce
1 medium fresh tomato
½ cup vegetable juice

Fats

Margarine, mayonnaise, vegetable oils, salad dressings, olives, and nuts. One exchange of fat is approximately 40 calories. Examples of 1 Fat choice or exchange:

1 teaspoon margarine or 2 teaspoons reduced-calorie margarine
1 teaspoon butter
1 teaspoon vegetable oil
1 teaspoon mayonnaise or 2 teaspoons reduced-calorie mayonnaise
1 teaspoon peanut butter
1 ounce olives
¼ ounce pecans or walnuts

Free Foods

Foods that do not provide nutritional value but are used to enhance the taste of foods are included in the Free Foods group. Examples of these are spices, herbs, extracts, vinegar, lemon juice, mustard, Worcestershire sauce, and soy sauce. Cooking sprays and artificial sweeteners used in moderation are also included in this group. However, you'll see that I include the caloric value of artificial sweeteners in the Optional Calories of the recipes.

You may occasionally see a recipe that lists "free food" as part of the portion. According to the published exchange lists, a free food contains fewer than 20 calories per serving. Two or three servings per day of free foods/drinks are usually allowed in a meal plan.

Optional Calories

Foods that do not fit into any other group but are used in moderation in recipes are included in Optional Calories. Foods that are counted in this way include sugar-free gelatin and puddings, fat-free mayonnaise and dressings, reduced-calorie whipped toppings, reduced-calorie syrups and jams, chocolate chips, coconut, and canned broth.

Sliders™

These are 80 Optional Calorie increments that do not fit into any particular category. You can choose which food group to *slide* these into. It is wise to limit this selection to approximately three to four per day to ensure the best possible nutrition for your body while still enjoying an occasional treat.

Sliders may be used in either of the following ways:

1. If you have consumed all your Protein, Bread, Fruit, or Skim Milk Weight Loss Choices for the day, and you want to eat additional foods from those food groups, you simply use a Slider. It's what I call "healthy horse trading." Remember that Sliders may not be traded for choices in the Vegetables or Fats food groups.

2. Sliders may also be deducted from your Optional Calories for the day or week. ¼ Slider equals 20 Optional Calories; ½ Slider equals 40 Optional Calories; ¾ Slider equals

60 Optional Calories; and 1 Slider equals 80 Optional Calories.

Healthy Exchanges Weight Loss Choices

My original Healthy Exchanges program of Weight Loss Choices was based on an average daily total of 1,400 to 1,600 calories per day. That was what I determined was right for my needs, and for those of most women. Because men require additional calories (about 1,600 to 1,900), here are my suggested plans for women and men. (*If you require more or fewer calories, please revise this plan to meet your individual needs.*)

Each day, women should plan to eat:

2 Skim Milk servings, 90 calories each
2 Fat servings, 40 calories each
3 Fruit servings, 60 calories each
4 Vegetable servings or more, 30 calories each
5 Protein servings, 60 calories each
5 Bread servings, 80 calories each

Each day, men should plan to eat:

2 Skim Milk servings, 90 calories each
4 Fat servings, 40 calories each
3 Fruit servings, 60 calories each
4 Vegetable servings or more, 30 calories each
6 Protein servings, 60 calories each
7 Bread servings, 80 calories each

Young people should follow the program for Men but add 1 Skim Milk serving for a total of 3 servings.

You may also choose to add up to 100 Optional Calories per day, and up to 21 to 28 Sliders per week at 80 calories each. If you choose to include more Sliders in your daily or weekly totals, deduct those 80 calories from your Optional Calorie "bank."

A word about **Sliders**: These are to be counted toward your totals after you have used your allotment of choices of Skim Milk, Protein, Bread, and Fruit for the day. By "sliding" an additional choice into one of these groups, you can meet your individual needs for that day. Sliders are especially helpful when traveling, stressed-out, eating out, or for special events. I often use mine so I can enjoy my favorite Healthy Exchanges desserts. Vegetables are not to be counted as Sliders. Enjoy as many Vegetable Choices as you need to feel satisfied. Because we want to limit our fat intake to moderate amounts, additional Fat Choices should not be counted as Sliders. If you choose to include more fat on an *occasional* basis, count the extra choices as Optional Calories.

Keep a daily food diary of your Weight Loss Choices, checking off what you eat as you go. If, at the end of the day, your required selections are not 100 percent accounted for, but you have done the best you can, go to bed with a clear conscience. There will be days when you have ¼ Fruit or ½ Bread left over. What are you going to do—eat two slices of an orange or half a slice of bread and throw the rest out? I always say that "Nothing in life comes out exact." Just do the best you can . . . *the best you can.*

Try to drink at least eight 8-ounce glasses of water a day. Water truly is the "nectar" of good health.

As a little added insurance, I take a multivitamin each day. It's not essential, but if my day's worth of well-planned meals "bites the dust" when unexpected events intrude on my regular routine, my body still gets its vital nutrients.

The calories listed in each group of Choices are averages. Some choices within each group may be higher or lower, so it's important to select a variety of different foods instead of eating the same three or four all the time.

Use your Optional Calories! They are what I call "life's little extras." They make all the difference in how you enjoy your food and appreciate the variety available to you. Yes, we can get by without them, but do you really want to? Keep in mind that you should be using all your daily Weight Loss Choices first to ensure you are getting the basics of good nutrition. But I guarantee that Optional Calories will keep you from feeling deprived—and help you reach your weight-loss goals.

Sodium, Fat, Cholesterol, and Processed Foods

re Healthy Exchanges Ingredients Really Healthy?
When I first created Healthy Exchanges, many people asked about sodium, about whether it was necessary to calculate the percentage of fat, saturated fat, and cholesterol in a healthy diet, and about my use of processed foods in many recipes. I researched these questions as I was developing my program, so you can feel confident about using the recipes and food plan.

Sodium

Most people consume more sodium than their bodies need. The American Heart Association and the American Diabetes Association recommend limiting daily sodium intake to no more than 3000 mg per day. If your doctor suggests you limit your sodium even more, then *you really must read labels.*

Sodium is an essential nutrient and should not be completely eliminated. It helps to regulate blood volume and is needed for normal daily muscle and nerve functions. Most of us, however, have no trouble getting "all we need" and then some.

As with everything else, moderation is my approach. I rarely ever have salt in my list as an added ingredient. But if you're especially sodium-sensitive, make the right choices for you—and save high-sodium foods such as sauerkraut for an occasional treat.

I use lots of spices to enhance flavors, so you won't notice the absence of salt. In the few cases where it is used, salt is vital for the success of the recipe, so please don't omit it.

When I do use an ingredient high in sodium, I try to compensate by using low-sodium products in the remainder of the recipe. Many fat-free products are a little higher in sodium to make up for any loss of flavor that disappeared along with the fat. But when I take advantage of these fat-free, higher-sodium products, I stretch that ingredient within the recipe, lowering the amount of sodium per serving. A good example is my use of fat-free and reduced-sodium canned soups. While the suggested number of servings per can is two, I make sure my final creation serves at least four and sometimes six. So the soup's sodium has been "watered down" from one-third to one-half of the original amount.

Even if you don't have to watch your sodium intake for medical reasons, using moderation is another "healthy exchange" to make on your own journey to good health.

Fat Percentages

We've been told that 30 percent is the magic number—that we should limit fat intake to 30 percent or less of our total calories. It's good advice, and I try to have a weekly average of 15 to 25 percent myself. I believe any less than 15 percent is really just another restrictive diet that won't last. And more than 25 percent on a regular basis is too much of a good thing.

When I started listing fat grams along with calories in my recipes, I was tempted to include the percentage of calories from fat. After all, in the vast majority of my recipes, that percentage is well below 30 percent This even includes my pie recipes that allow you a realistic serving instead of many "diet" recipes that tell you a serving is $1/12$ of a pie.

Figuring fat grams is easy enough. Each gram of fat equals 9 calories. Multiply fat grams by 9, then divide that number by the total calories to get the percentage of calories from fat.

So why don't I do it? After consulting with four registered dietitians for advice, I decided to omit this information. They felt that it's too easy for people to become obsessed by that 30 percent figure, which is after all supposed to be a percentage of total calories over the course of a day or a week. We mustn't feel we can't include a healthy ingredient such as pecans or olives in one recipe just because, on its own, it has more than 30 percent of its calories from fat.

An example of this would be a casserole made with 90 percent lean red meat. Most of us benefit from eating red meat in moderation, as it provides iron and niacin in our diets, and it also makes life more enjoyable for us and those who eat with us. If we *only* look at the percentage of calories from fat in a serving of this one dish, which might be as high as 40 to 45 percent, we might choose not to include this recipe in our weekly food plan.

The dietitians suggested that it's important to consider the total picture when making such decisions. As long as your overall food plan keeps fat calories to 30 percent, it's all right to enjoy an occasional dish that is somewhat higher in fat content. Healthy foods I include in **MODERATION** include 90 percent lean red meat, olives, and nuts. I don't eat these foods every day, and you may not either. But occasionally, in a good recipe, they make all the difference in the world between just getting by (deprivation) and truly enjoying your food.

Remember, the goal is eating in a healthy way so you can enjoy and live well the rest of your life.

Saturated Fats and Cholesterol

You'll see that I don't provide calculations for saturated fats or cholesterol amounts in my recipes. It's for the simple and yet not so simple reason that accurate, up-to-date, brand-specific information can be difficult to obtain from food manufacturers, especially since the way in which they produce food keeps changing rapidly. But once more I've consulted with registered dietitians and other professionals and found that, because I use only a few products that are high in saturated fat, and use them in such limited quantities, my recipes are suitable for patients concerned about controlling or lowering cholesterol. You'll also find that whenever I do use one of these ingredients *in moderation*, everything else in the recipe, and in the meals my family and I enjoy, is low in fat.

Processed Foods

Some people have asked how "healthy" recipes can so often use "processed foods"—ready-made products like canned soups, prepared pie crusts, frozen potatoes, and frozen whipped topping? Well, I believe that such foods, used properly (that word **moderation** again) as part of a healthy lifestyle, have a place as ingredients in healthy recipes.

I'm not in favor of spraying everything we eat with chemicals, and I don't mean that all our foods should come out of packages. But I do think we should use the best available products to make cooking easier and foods taste better. I take advantage of good low-fat and low-sugar products, and my recipes are created for busy people like me who want to eat well and eat healthy. I don't expect people to visit out-of-the-way health food stores or find time to cook beans from scratch—*because I don't*. There are lots of very good processed foods available in your local grocery store, and they can make it so much easier to enjoy the benefits of healthy eating.

I certainly don't recommend that everything you eat come from a can, box, or jar. I think the best of all possible worlds is to start with the basics: rice, poultry, fish, or beef, and raw vegetables—then throw in a can of reduced-sodium/97 percent fat-free soup (a processed food) and end up with an appetizing, easy-to-prepare healthy meal.

Most of us can't grow fresh food in the backyard, and many people don't even have a nearby farmer's market. But instead of saying, "Well, I can't get to the health food store so why not eat that hot fudge sundae?" you gotta play ball in your private ball field, not in someone else's. I want to help you figure out ways to make living healthy **doable** and **livable** *wherever you live,* or you're not going to stick with it.

I've checked with the American Dietetic Association, the

American Diabetic Association, and with many registered dietitians, and I've been assured that sugar-free and fat-free processed products that use sugar and fat substitutes are safe when used in the intended way. This means a realistic serving, not one hundred cans of diet soda every day of the year! Even carrots can turn your skin orange if you eat far too many, but does anyone suggest we avoid eating carrots?

Of course, it is your privilege to disagree with me and to use whatever you choose when you prepare your food. I never want to be one of those "opinionated" people who think it's their God-given right to make personal decisions for others and insist that their way is the *only* way.

Besides, new research comes out every day that declares one food bad and another food good. Then a few days later, some new information emerges, saying that the opposite is true. When the facts are sifted from the fiction, the truth is probably somewhere in between. I know I feel confused when what was bad for you last year is good for you now, and vice versa.

Instead of listening to unreasonable sermons by naysayers who are nowhere around when it comes time to make a quick and healthy meal for your family, I've tried to incorporate the best processed foods I can find into my Healthy Exchanges recipes. I get stacks of mail from people who are thrilled to discover they can eat good-tasting food and who proudly use processed foods in the intended way. I think you will agree that my commonsense approach to healthy cooking is the right choice for many. Because these foods are convenient, tasty, and good substitutes for less healthy products, people are willing to use them long-term.

So don't let anyone make you feel ashamed for including these products in your healthy lifestyle. Only you can decide what's best for you and your family's needs. Part of living a healthy lifestyle is making those decisions and *getting on with life.*

My Best Healthy Exchanges Tips and Tidbits

Measurements, General Cooking Tips, and Basic Ingredients

The word **moderation** best describes **my use of fats, sugar substitutes,** and **sodium** in these recipes. Wherever possible, I've used cooking spray for sautéing and for browning meats and vegetables. I also use reduced-calorie margarine and no-fat mayonnaise and salad dressings. Lean ground turkey *or* ground beef can be used in the recipes. Just be sure whatever you choose is at least *90 percent lean.*

I've also included **small amounts of sugar and brown sugar substitutes as the sweetening agent** in many of the recipes. I don't drink a hundred cans of soda a day or eat enough artificially sweetened foods in a 24-hour time period to be troubled by sugar substitutes. But if this is a concern of yours and you *do not* need to watch your sugar intake, you can always replace the sugar substitutes with processed sugar and the sugar-free products with regular ones.

I created my recipes knowing they would also be used by hypoglycemics, diabetics, and those concerned about triglycerides. If you choose to use sugar instead, be sure to count the additional calories.

A word of caution when cooking with **sugar substitutes**: Use **saccharin**-based sweeteners when **heating or baking**. In recipes that **don't require heat, Aspartame** (known as Nutrasweet) works well in uncooked dishes but leaves an aftertaste in baked products.

I'm often asked why I use an **8-by-8-inch baking dish** in my recipes. It's for portion control. If the recipe says it serves 4, just cut down the center, turn the dish, and cut again. Like magic, there's your serving. Also, if this is the only recipe you are preparing requiring an oven, the square dish fits into a tabletop toaster oven easily and energy can be conserved.

To make life even easier, **whenever a recipe calls for ounce measurements** (other than raw meats) I've included the closest cup equivalent. I need to use my scale daily when creating recipes, so I've measured for you at the same time.

Most of the recipes are for **4 to 6 servings.** If you don't have that many to feed, do what I do: freeze individual portions. Then all you have to do is choose something from the freezer and take it to work for lunch or have your evening meals prepared in advance for the week. In this way, I always have something on hand that is both good to eat and good for me.

Unless a recipe includes hard-cooked eggs, cream cheese, mayonnaise, or a raw vegetable or fruit, **the leftovers should freeze well**. (I've marked recipes that freeze well with the symbol of a **snowflake ❄.**)This includes most of the cream pies. Divide any recipe up into individual servings and freeze for your own "TV" dinners.

Another good idea is **cutting leftover pie into individual pieces and freezing each one separately** in a small Ziploc freezer bag. Then the next time you want to thaw a piece of pie for yourself, you don't have to thaw the whole pie. It's great this way for brown-bag lunches, too. Just pull a piece out of the freezer on your way to work and by lunchtime you will have a wonderful dessert waiting for you.

Unless I specify **"covered" for simmering or baking,** prepare my recipes **uncovered**. Occasionally you will read a recipe that asks you to cover a dish for a time, then to uncover, so read the directions carefully to avoid confusion—and to get the best results.

Low-fat cooking spray is another blessing in a Healthy Exchanges kitchen. It's currently available in three flavors . . .

- OLIVE OIL–FLAVORED when cooking Mexican, Italian, or Greek dishes

- BUTTER FLAVORED when the hint of butter is desired

- REGULAR for everything else.

A quick spray of butter flavored makes air-popped popcorn a low-fat taste treat, or try it as a butter substitute on steaming hot corn on the cob. One light spray of the skillet when browning meat will convince you that you're using "old-fashioned fat," and a quick coating of the casserole dish before you add the ingredients will make serving easier and cleanup quicker.

I use reduced-sodium **canned chicken broth** in place of dry bouillon to lower the sodium content. The intended flavor is still present in the prepared dish. As a reduced-sodium beef broth is not currently available (at least not in DeWitt, Iowa), I use the canned regular beef broth. The sodium content is still lower than regular dry bouillon.

Whenever **cooked rice or pasta** is an ingredient, follow the package directions, but eliminate the salt and/or margarine called for. This helps lower the sodium and fat content. It tastes just fine; trust me on this.

Here's another tip: When **cooking rice or noodles**, why not cook extra "for the pot"? After you use what you need, store leftover rice in a covered container (where it will keep for a couple of days). With noodles like spaghetti or macaroni, first rinse and drain as usual, then measure out what you need. Put the leftovers in a bowl covered with water, then store in the refrigerator, covered, until they're needed. Then, measure out what you need, rinse and drain them, and they're ready to go.

Does your **pita bread** often tear before you can make a sandwich? Here's my tip to make them open easily: cut the bread in half, put the halves in the microwave for about 15 seconds, and they will open up by themselves. *Voilà!*

When **chunky salsa** is listed as an ingredient, I leave the degree of "heat" up to your personal taste. In our house, I'm considered a wimp. I go for the "mild" while Cliff prefers "extra-hot." How do we compromise? I prepare the recipe with mild salsa because he can always add a spoonful or two of the hotter version to his serving, but I can't enjoy the dish if it's too spicy for me.

Proteins

I use eggs in moderation. I enjoy the real thing on an average of three to four times a week. So, my recipes are calculated on using whole eggs. However, if you choose to use egg substitute in place of the egg, the finished product will turn out just fine and the fat grams per serving will be even lower than those listed.

If you like the look, taste, and feel of **hard-boiled eggs** in salads, but haven't been using them because of the cholesterol in the yolk, I have a couple of alternatives for you. 1) Pour an 8-oz carton of egg substitute into a medium skillet sprayed with cooking spray. Cover skillet tightly and cook over low heat until substitute is just set, about 10 minutes. Remove from heat and let set, still covered, for 10 minutes more. Uncover and cool completely. Chop set mixture. This will make about 1 cup of chopped egg. 2) Even easier is to hard-boil "real eggs," toss the yolk away, and chop the white. Either way, you don't deprive yourself of the pleasure of egg in your salad.

In most recipes calling for **egg substitutes**, you can use 2 egg whites in place of the equivalent of 1 egg substitute. Just break the eggs open and toss the yolks away. I can hear some of you already saying, "But that's wasteful!" Well, take a look at the price on the egg substitute package (which usually has the equivalent of 4 eggs in it), then look at the price of a dozen eggs, from which you'd get the equivalent of 6 egg substitutes. Now, what's wasteful about that?

Whenever I include **cooked chicken** in a recipe, I use roasted white meat without skin. Whenever I include **roast beef or pork** in a recipe, I use the loin cuts because they are much leaner. However, most of the time, I do my roasting of all these meats at the local deli.

I just ask for a chunk of their lean roasted meat, 6 or 8 ounces, and ask them not to slice it. When I get home, I cube or dice the meat and am ready to use it in my recipe. The reason I do this is three-fold: 1) I'm getting just the amount I need without leftovers; 2) I don't have the expense of heating the oven; 3) I'm not throwing away the bone, gristle, and fat I'd be cutting away from the meat. Overall, it is probably cheaper to "roast" it the way I do.

Did you know that you can make an acceptable meat loaf without using egg for the binding? Just replace every egg with ¼ cup of liquid. You could use beef broth, tomato sauce, even apple-sauce, to name just a few. For a meat loaf to serve 6, I always use 1 pound of extra-lean ground beef or turkey, 6 tablespoons of dried fine bread crumbs, and ¼ cup of the liquid, plus anything else healthy that strikes my fancy at the time. I mix well and place the mixture in an 8-by-8-inch baking dish or 9-by-5-inch loaf pan sprayed with cooking spray. Bake uncovered at 350 degrees for 35 to 50 minutes (depending on the added ingredients). You will never miss the egg.

Any time you are **browning ground meat** for a casserole and want to get rid of almost all the excess fat, just place the uncooked meat loosely in a plastic colander. Set the colander in a glass pie plate. Place in microwave and cook on HIGH for 3 to 6 minutes (depending on the amount being browned), stirring often. Use as you would for any casserole. You can also chop up onions and brown them with the meat if you want.

Fruits and Vegetables

If you want to enjoy a **"fruit shake"** with some pizazz, just combine soda water and unsweetened fruit juice in a blender. Add crushed ice. Blend on HIGH until thick. Refreshment without guilt.

You'll see that many recipes use ordinary **canned vegetables.** They're much cheaper than reduced-sodium versions, and once you rinse and drain them, the sodium is reduced anyway. I believe in saving money wherever possible so we can afford the best fat-free and sugar-free products as they come onto the market.

All three kinds of **vegetables—fresh, frozen, and canned—** have their place in a healthy diet. My husband, Cliff, hates the taste of frozen or fresh green beans, thinks the texture is all wrong, so I use canned green beans instead. In this case, canned vegetables have their proper place when I'm feeding my husband. If someone in your family has a similar concern, it's important to respond to it so everyone can be happy and enjoy the meal.

When I use **fruits or vegetables** like apples, cucumbers, and zucchini, I wash them really well and **leave the skin on.** It provides added color, fiber, and attractiveness to any dish. And, because I use processed flour in my cooking, I like to increase the fiber in my diet by eating my fruits and vegetables in their closest-to-natural state.

To help keep **fresh fruits and veggies fresh**, just give them a quick "shower" with lemon juice. The easiest way to do this is to pour purchased lemon juice into a kitchen spray bottle and store in the refrigerator. Then, every time you use fresh fruits or vegetables in a salad or dessert, simply give them a quick spray with your "lemon spritzer." You just might be amazed by how this little trick keeps your produce from turning brown so fast.

The next time you warm canned vegetables such as carrots or green beans, drain and heat the vegetables in $\frac{1}{4}$ cup beef or chicken broth. It gives a nice variation to an old standby. Here's a simple **white sauce** for vegetables and casseroles without using added fat that can be made by spraying a medium saucepan with butter-flavored cooking spray. Place $1\frac{1}{2}$ cups evaporated skim milk and 3 tablespoons flour in a covered jar. Shake well. Pour into sprayed saucepan and cook over medium heat until thick, stirring constantly. Add salt and pepper to taste. You can also add $\frac{1}{2}$ cup canned drained mushrooms and/or 3 ounces ($\frac{3}{4}$ cup) shredded reduced-fat cheese. Continue cooking until cheese melts.

Zip up canned or frozen green beans with **chunky salsa**: $\frac{1}{2}$ cup to 2 cups beans. Heat thoroughly. Chunky salsa also makes a wonderful dressing on lettuce salads. It only counts as a vegetable, so enjoy.

Another wonderful **South of the Border** dressing can be stirred up by using $\frac{1}{2}$ cup of chunky salsa and $\frac{1}{4}$ cup fat-free Ranch dressing. Cover and store in your refrigerator. Use as a dressing for salads or as a topping for baked potatoes.

For **gravy** with all the "old time" flavor but without the extra fat, try this almost effortless way to prepare it. (It's almost as easy as opening up a store-bought jar.) Pour the juice off your roasted meat, then set the roast aside to "rest" for about 20 minutes. Place the juice in an uncovered cake pan or other large flat pan (we want the large air surface to speed up the cooling process) and put in the freezer until the fat congeals on top and you can skim it off. Or, if you prefer, use a skimming pitcher purchased at your kitchen gadget store. Either way, measure about 1½ cups skimmed broth and pour into a medium saucepan. Cook over medium heat until heated through, about 5 minutes. In a covered jar, combine ½ cup water or cooled potato broth with 3 tablespoons flour. Shake well. Pour flour mixture into warmed juice. Combine well using a wire whisk. Continue cooking until gravy thickens, about 5 minutes. Season with salt and pepper to taste.

Why did I use flour instead of cornstarch? Because any leftovers will reheat nicely with the flour base and would not with a cornstarch base. Also, 3 tablespoons of flour works out to 1 Bread/Starch exchange. This virtually fat-free gravy makes about 2 cups, so you could spoon about ½ cup gravy on your low-fat mashed potatoes and only have to count your gravy as ¼ Bread/Starch exchange.

Desserts

Thaw **lite whipped topping** in the refrigerator overnight. Never try to force the thawing by stirring or using a microwave to soften. Stirring it will remove the air from the topping that gives it the lightness and texture we want, and there's not enough fat in it to survive being heated.

How can I **frost an entire pie with just ½ cup of whipped topping?** First, don't use an inexpensive brand. I use Cool Whip Lite or La Creme Lite. Make sure the topping is fully thawed. Always spread from the center to the sides using a rubber spatula. This way, ½ cup topping will literally cover an entire pie. Remember, the operative word is *frost*, not pile the entire container on top of the pie!

For a special treat that tastes anything but "diet," try placing **spreadable fruit** in a container and microwave for about 15 seconds. Then pour the melted fruit spread over a serving of nonfat ice cream or frozen yogurt. One tablespoon of spreadable fruit is equal to 1 fruit serving. Some combinations to get you started are apricot over chocolate ice cream, strawberry over strawberry ice cream, or any flavor over vanilla. Another way I use spreadable fruit is to make a delicious **topping for a cheesecake or angel food cake**. I take ½ cup of fruit and ½ cup Cool Whip Lite and blend the two together with a teaspoon of coconut extract.

Here's a really **good topping** for the fall of the year. Place 1½ cups unsweetened applesauce in a medium saucepan or 4-cup glass measure. Stir in 2 tablespoons raisins, 1 teaspoon apple pie spice, and 2 tablespoons Cary's Sugar-Free Maple Syrup. Cook over medium heat on stove or on HIGH in microwave until warm. Then spoon about ½ cup warm mixture over pancakes, french toast, or fat-free and sugar-free vanilla ice cream. It's as close as you will get to guilt-free apple pie!

A quick yet tasty way to prepare **strawberries for shortcake** is to place about ¾ cup sliced strawberries, 2 tablespoons Diet Mountain Dew, and sugar substitute to equal ¼ cup sugar in a blender container. Process on BLEND until mixture is smooth. Pour mixture into bowl. Add 1¼ cups sliced strawberries and mix well. Cover and refrigerate until ready to serve with shortcake.

The next time you are making treats for the family, try using **unsweetened applesauce** for some or all of the required oil in the recipe. For instance, if the recipe calls for ½ cup cooking oil, use up to ½ cup in applesauce. It works and most people will not even notice the difference. It's great in purchased cake mixes, but so far I haven't been able to figure out a way to deep-fat fry with it!

Another trick I often use is to include tiny amounts of "real people" food, such as coconut, but extend the flavor by using extracts. Try it—you will be surprised by how little of the real thing you can use and still feel you are not being deprived.

If you are preparing a pie filling that has ample moisture, just line **graham crackers** in the bottom of a 9-by-9-inch cake pan. Pour the filling over the top of the crackers. Cover and refrigerate until the moisture has enough time to soften the crackers. Overnight is best. This eliminates the added **fats and sugars of a pie crust.**

When **stirring fat-free cream cheese to soften it,** use only a sturdy spoon, never an electric mixer. The speed of a mixer can cause the cream cheese to lose its texture and become watery.

Did you know you can make your own **fruit-flavored yogurt?** Mix 1 tablespoon of any flavor of spreadable fruit spread with ¾ cup plain yogurt. It's every bit as tasty and much cheaper. You can also make your own **lemon yogurt** by combining 3 cups plain fat-free yogurt with 1 tub Crystal Light lemonade powder. Mix well, cover, and store in refrigerator. I think you will be pleasantly surprised by the ease, cost, and flavor of this "made from scratch" calcium-rich treat. P.S.: You can make any flavor you like by using any of the Crystal Light mixes—Cranberry? Iced tea? You decide.

Sugar-free puddings and gelatins are important to many of my recipes, but if you prefer to avoid sugar substitutes, you could still prepare the recipes with regular puddings or gelatins. The calories would be higher, but you would still be cooking low-fat.

When a recipe calls for **chopped nuts** (and you only have whole ones), who wants to dirty the food processor just for a couple of tablespoons? You could try to chop them using your cutting board, but be prepared for bits and pieces to fly all over the kitchen. I use "Grandma's food processor." I use the biggest nuts I can find, put them in a small glass bowl, and chop them into chunks just the right size using a metal biscuit cutter.

If you have a **leftover muffin** and are looking for something a little different for breakfast, you can make **a "breakfast sundae."** Crumble the muffin into a cereal bowl. Sprinkle a serving of fresh fruit over it and top with a couple tablespoons nonfat plain yogurt sweetened with sugar substitute and your choice of extract. The thought of it just might make you jump out of bed with a smile on your face. (Speaking of muffins, did you know that if you fill the unused muffin wells with water when baking muffins, you help ensure more even baking and protect the muffin pan at the same time?) Another muffin hint: Lightly spray the inside of paper baking cups with butter-flavored cooking spray before spooning the muffin batter into them. Then you won't end up with paper clinging to your fresh-baked muffins.

The secret of making **good meringues** without sugar is to use 1 tablespoon of Sprinkle Sweet or Sugar Twin for every egg white, and a small amount of extract. Use ½ to 1 teaspoon for the batch. Almond, vanilla, and coconut are all good choices. Use the same amount of cream of tartar you usually do. Bake the meringue in the same old way. Don't think you can't have meringue pies because you can't eat sugar. You can, if you do it my way. (Remember that egg whites whip up best at room temperature.)

Homemade or Store-Bought?

I've been asked which is better for you: homemade from scratch, or purchased foods. My answer is *both!* They each have a place in a healthy lifestyle, and what that place is has everything to do with you.

Take **piecrusts**, for instance. If you love spending your spare time in the kitchen preparing foods, and you're using low-fat, low-sugar, and reasonably low sodium ingredients, go for it! But if, like so many people, your time is limited and you've learned to read labels, you could be better off using purchased foods.

I know that when I prepare a pie (and I experiment with a couple of pies each week, because this is Cliff's favorite dessert) I use a purchased crust. Why? Mainly because I can't make a good-tasting pie crust that is lower in fat than the brands I use. Also, purchased piecrusts fit my rule of "If it takes longer to fix than to eat, forget it!"

I've checked the nutrient information for purchased piecrusts against recipes for traditional and "diet" piecrusts, using my computer software program. The purchased crust calculated lower in both fat and calories! I have tried some low-fat and low-sugar recipes, but they just didn't spark my taste buds, or were so complicated you needed an engineering degree just to get the crust in the pie plate.

I'm very happy with the purchased piecrusts in my recipes, because the finished product rarely, if ever, has more than 30 percent of total calories coming from fats. I also believe that we have to prepare foods our families and friends will eat with us on a regular basis and not feel deprived, or we've wasted time, energy, and money.

I could use a purchased "lite" **pie filling**, but instead I make my own. Here I can save both fat and sugar, and still make the filling almost as fast as opening a can. The bottom line: Know what you have to spend when it comes to both time and fat/sugar calories, then make the best decision you can for you and your family.

And don't go without an occasional piece of pie because you think it isn't *necessary*. A delicious pie prepared in a healthy way is one of the simple pleasures of life. It's a little thing, but it can make all the difference between just getting by with the bare minimum and living a full and healthy lifestyle.

Many people have experimented with my tip about **substituting applesauce and artificial sweetener for butter and sugar**, but what if you aren't satisfied with the result? One woman wrote to me about a recipe for her grandmother's cookies that called for 1 cup butter and 1½ cups sugar. Well, any recipe that depends on so much butter and sugar as this one does is generally not a good candidate for "healthy exchanges." The original recipe needed a large quantity of fat to produce a crisp cookie just like Grandma made.

Unsweetened applesauce can be used to substitute for vegetable oil with various degrees of success, but not to replace butter, lard, or margarine. If your recipe calls for ½ cup oil or less, and it's a quick bread, muffin, or bar cookie, it should work to replace the oil with applesauce. If the recipe calls for more than ½ cup oil, then experiment with half oil, half applesauce. You've still made the recipe healthier, even if you haven't removed all the oil from it.

Another rule for healthy substitution: Up to ½ cup sugar or less can be replaced by *an artificial sweetener that can withstand the heat of baking*, like Sugar Twin or Sprinkle Sweet. If it requires more than ½ cup sugar, cut the amount needed by 75 percent and use ½ cup sugar substitute and sugar for the rest. Other options: Reduce the butter and sugar by 25 percent and see if the finished product still satisfies you in taste and appearance. Or, make the cookies just like Grandma did, realizing they are part of your family's holiday tradition. Enjoy a moderate serving of a couple of cookies once or twice during the season, and just forget about them the rest of the year.

I'm sure you'll add to this list of cooking tips as you begin preparing Healthy Exchanges recipes and discover how easy it can be to adapt your own favorite recipes using these ideas and your own common sense.

A Peek into
My Pantry and My
Favorite Brands

Everyone asks me what foods I keep on hand and what brands I use. There are lots of good products on the grocery shelves today—many more than we dreamed about even a year or two ago. And I can't wait to see what's out there twelve months from now. The following are my staples and, where appropriate, my favorites *at this time*. I feel these products are healthier, tastier, easy to get—and deliver the most flavor for the least amount of fat, sugar, or calories. If you find others you like as well *or better*, please use them. This is only a guide to make your grocery shopping and cooking easier.

Fat-free plain yogurt (*Yoplait*)

Nonfat dry skim milk powder (*Carnation*)

Evaporated skim milk (*Carnation*)

Skim milk

Fat-free cottage cheese

Fat-free cream cheese (*Philadelphia*)

Fat-free mayonnaise (*Kraft*)

Fat-free salad dressings (*Kraft*)

Fat-free sour cream (*Land O Lakes*)

Reduced-calorie margarine (*Weight Watchers, Promise*, or *Smart Beat*)

Cooking spray:

Olive oil–flavored and regular (*Pam*)

Butter flavored for sautéing (*Weight Watchers*)

Butter flavored for spritzing *after* cooking (*I Can't Believe It's Not Butter!*)

Vegetable oil (*Puritan Canola Oil*)

Reduced-calorie whipped topping (*Cool Whip Lite* or *Cool Whip Free*)

Sugar Substitute:

if no heating is involved (*Equal*)

if heating is required

white (*Sugar Twin* or *Sprinkle Sweet*)

brown (*Brown Sugar Twin*)

Sugar-free gelatin and pudding mixes (*JELL-O*)

Baking mix (*Bisquick Reduced Fat*)

Pancake mix (*Aunt Jemima Reduced-Calorie*)

Reduced-calorie pancake syrup (*Cary's Sugar-Free*)

Parmesan cheese (*Kraft fat-free* or *Weight Watchers fat-free*)

Reduced-fat cheese (*Kraft ⅓ Less Fat* and *Weight Watchers*)

Shredded frozen potatoes (*Mr. Dell's*)

Spreadable fruit spread (*Smucker's, Welch's* or *Sorrell Ridge*)

Peanut butter (*Peter Pan Reduced-Fat, Jif Reduced-Fat,* or *Skippy Reduced-Fat*)

Chicken broth (*Healthy Request*)

Beef broth (*Swanson*)

Tomato sauce (*Hunt's—Chunky* and *Regular*)

Canned soups (*Healthy Request*)

Tomato juice (*Campbell's Reduced-Sodium*)

Ketchup (*Heinz Lite Harvest* or *Healthy Choice*)

Purchased piecrust

unbaked (*Pillsbury—from dairy case*)

graham cracker, butter-flavored, or chocolate-flavored (*Keebler*)

Crescent rolls (*Pillsbury Reduced-Fat*)

Pastrami and corned beef (*Carl Buddig Lean*)

Luncheon meats (*Healthy Choice or Oscar Mayer*)

Ham (*Dubuque 97% Fat Free and Reduced Sodium* or *Healthy Choice*)

Frankfurters and Kielbasa sausage (*Healthy Choice*)

Canned white chicken, packed in water (*Swanson*)

Canned tuna, packed in water (*Starkist*)

90 percent lean ground turkey and beef

Soda crackers (*Nabisco Fat-Free*)

Reduced-calorie bread—40 calories per slice or less

Hamburger buns—80 calories each (*Colonial Old Fashion* or *Less*)

Rice—instant, regular, brown, and wild

Instant potato flakes (*Betty Crocker Potato Buds*)

Noodles, spaghetti, and macaroni

Salsa (*Chi Chi's Mild*)

Pickle relish—dill, sweet, and hot dog

Mustard—Dijon, prepared, and spicy

Unsweetened apple juice

Unsweetened applesauce

Fruit—fresh, frozen (no sugar added), or canned in juice

Vegetables—fresh, frozen, or canned

Spices—JO's Spices

Lemon and lime juice (in small plastic fruit-shaped bottles found in produce section)

Instant fruit beverage mixes (*Crystal Light*)

Dry dairy beverage mixes (*Nestlé's Quik* and *Swiss Miss*)

"Ice Cream" (*Well's Blue Bunny Health Beat Fat* and *Sugar Free*)

The items on my shopping list are everyday foods found in just about any grocery store in America. But all are as low in fat, sugar, calories, and sodium as I can find—and they still taste good! I can make any recipe in my cookbooks and newsletters as long as I have my cupboards and refrigerator stocked with these items. Whenever I use the last of any one item, I just make sure I pick up another supply the next time I'm at the store.

If your grocer does not stock these items, why not ask if they can be ordered on a trial basis? If the store agrees to do so, be sure to tell your friends to stop by, so that sales are good enough to warrant restocking the new products. Competition for shelf space is fierce, so only products that sell well stay around.

The Recipes

JoAnna's Welcome to the Kitchen

Welcome to the Healthy Exchanges Test Kitchen, the home of easy "common folk" healthy recipes. During our time together, I'll be sharing some of the best tips you will ever come across for creating your own healthy dairy products at home simply by starting with a box of dry milk powder. Even people who vowed never to touch a box of that "white powder" will become converts! (In fact, once these hints hit the streets, I'm sure the price of nonfat dry milk powder stock will go through the roof, so you may want to contact a stockbroker NOW if you are into that sort of thing. . . .)

Of all the products in the grocery store that "get no respect," dry milk powder has to top the list. Many of us can share horror stories of drinking "that funny-tasting stuff" when we were younger and promising ourselves we would never drink it again once we were grown. I know; I felt the same way myself.

But when I got tired of tossing away leftover evaporated skim milk because so many recipes (not mine, I'm proud to proclaim) just never used the whole can, I discovered the magic of dry milk powder. Before that, I'd dutifully put the evaporated skim milk leftovers in the refrigerator with lots of good intentions to use up the last few ounces in the next day or two. More often than not, however, when I got around to cleaning out the refrigerator a few months later, I'd inevitably have to toss out many an open can of spoiled evaporated skim milk!

I knew there had to be a better way, and that's when I began experimenting with just enough dry milk powder and water to stir up the amount of evaporated milk I needed for a recipe. When this produced wonderful results, I decided to see what else I could do. That's when I discovered all the dairy products I could create using my "magical" powder. I quickly became a FIRST CLASS FAN of that "calcium-in-a-box"! Five years ago, I wouldn't have been caught with a box in the house; now, I wouldn't dream of ever being without it!

Don't give nonfat dry milk powder a "bum steer." It's great! I *do not* use it for drinking, but I *do* use it extensively for cooking. If I had to choose only one product I would never be without, it would be nonfat dry milk powder. I've said many a time, "Give me my mixing bowl, wire whisk, and dry milk powder—and I can conquer the world!"

Three of the many reasons why I love to use nonfat dry milk powder for cooking are:

1. **It is very inexpensive.** If you compare the cost of the number of quarts of skim milk you would need to purchase to match the number of quarts your box of dry milk powder will make, you will discover immediately that this precious box is indeed a friend of the "penny pincher."

2. **It does not sour** within the week like fresh milk. If you don't use dry milk powder on a regular basis, then store the box in your refrigerator or freezer and its shelf life will be almost forever! Yes, dry milk powder can become rancid if you buy a box, plop it in your cupboard, and forget about it for months or years on end. Just remember that a box of dry milk powder will stay fresh on the shelf for at least three months, and much longer if you put it "on ice."

3. **It is a very easy way to add extra calcium** to just about any recipe without adding liquid. Add calcium without fuss to any family favorite by stirring dry milk powder into casseroles, quick breads, or desserts. Try it, and feel your bones say "thank you!" for the boost in calcium.

Take a look through this section to see the many, many ways I make my own healthy, lowfat dairy products with dry milk powder. After trying my recipes, you can quickly start to use my **JO's Dairy Mixes** in your own recipes. You, too, might just become an advocate for nonfat dry milk powder. Remember my slogan: "Nonfat Dry Milk Powder. . .Mother Nature's Modern-Day Convenience."

But, just because I think dry milk powder is the greatest invention since sliced bread doesn't mean I don't use other calcium-rich foods, because I do. In this collection of recipes, you will find creamy chowders using reduced-fat cheeses, delectable salads stirred up with fat-free sour cream, and scrumptious cheesecakes made with fat-free cream cheese. You will also find an abundance of vegetable dishes so full of flavor that they are guaranteed to win you praise from your loved ones. It's up to you whether you let on that they are easy to prepare and oh-so-healthy!

When I began working on this book, I set a goal for myself: I wanted every single recipe to provide at least 76 milligrams of calcium per serving, the equivlent of one-quarter cup skim milk. I'm happy to say I succeeded, and you'll see that many recipes deliver even more calcium than that!

How to Read a Healthy Exchanges Recipe

The Healthy Exchanges Nutritional Analysis

Before using these recipes you may wish to consult your physician or health-care provider to be sure they are appropriate for you. The information in this book is not intended to take the place of any medical advice. It reflects my experiences, studies, research, and opinions regarding healthy eating.

Each recipe includes nutritional information calculated in three ways:

> Healthy Exchanges Weight-Loss Choices/Exchanges
> Calories, calcium, fiber, and fat grams
> Diabetic exchanges

In every Healthy Exchanges recipe, the diabetic exchanges have been calculated by a registered dietitian. All the other calculations were done by computer, using the Food Processor II software. When the ingredient listing gives more than one choice, the first ingredient listed is the one used in the recipe analysis. Due to inevitable variations in the ingredients you choose to use, the nutritional values should be considered approximate.

The annotation "(limited)" following Protein counts in some recipes indicates that consumption of whole eggs should be limited to four per week.

Please note the following symbols:

☆ This star means read the recipe's directions carefully for special instructions about **division** of ingredients.

❋ This symbol indicates **FREEZES WELL**.

JO's Dairy Mixes

Many of the recipes in this book call for special mixes I've created and use constantly. Rather than repeat these mix recipes every time, I've put them in this section for easy reference. You can also use these mixes in your own recipes when you want to substitute low-fat and low-sugar versions for the standards.

JO's Evaporated Skim Milk

⅓ cup Carnation Nonfat Dry Milk Powder
½ cup water

In a small bowl, combine dry milk powder and water. Mix well. Use as you would purchased evaporated skim milk. Makes ½ cup.

The entire recipe equals:

HE: 1 Skim Milk

---

80 Calories • 0 gm Fat • 8 gm Protein •
12 gm Carbohydrate • 123 mg Sodium •
276 mg Calcium • 0 gm Fiber

DIABETIC: 1 Skim Milk

HINT: If you need 1 cup evaporated skim milk, double the recipe. For 1½ cups, which is equivalent to one 12-fluid-ounce can, simply triple the recipe. This is handy to know when you want to prepare a recipe calling for evaporated skim milk and you don't have any in the cupboard. Also, if you are using a recipe that only needs ½ cup evaporated skim milk, you don't have to worry about what to do with the leftovers remaining in the can.

JO's Buttermilk

⅔ cup Carnation Nonfat Dry Milk Powder

1 cup water

2 teaspoons white vinegar

In a small bowl, combine dry milk powder and water. Stir in vinegar. Let set 10 minutes. Use in any recipe that calls for buttermilk. Makes 1 cup.

The entire recipe equals:

HE: 2 Skim Milk

---

160 Calories • 0 gm Fat • 16 gm Protein •
24 gm Carbohydrate • 246 mg Sodium •
552 mg Calcium • 0 gm Fiber

DIABETIC: 2 Skim Milk

Yes, you can make "sour milk" by adding vinegar to regular milk. But, the texture is much thinner than buttermilk and doesn't work out quite as well in recipes. My version is thicker, so the final product turns out more like you used real buttermilk.

JO's Sour Cream

¾ cup Yoplait plain fat-free yogurt
⅓ cup Carnation Nonfat Dry Milk Powder

In a small bowl, combine yogurt and dry milk powder. Use as you would purchased sour cream. Makes 1 cup.

The entire recipe equals:

HE: 2 Skim Milk

---

172 Calories • 0 gm Fat • 18 gm Protein •
25 gm Carbohydrate • 251 mg Sodium •
276 mg Calcium • 0 gm Fiber

DIABETIC: 2 Skim Milk

What you are doing is four-fold: (1) the dry milk stabilizes the yogurt and keeps the whey from separating; (2) the dry milk powder slightly helps to cut the tartness of the yogurt; (3) it's still virtually fat-free; (4) and the calcium has been increased by 100 percent.

Use "as is" in any recipe that doesn't require heat. If you want to use in a stroganoff or a quick bread, simply stir 1 teaspoon cornstarch into **JO's Sour Cream** before adding it to the other ingredients, as this keeps the yogurt from separating.

Isn't it great how we can make yogurt, that distant relative of sour cream, a first kissin' cousin by adding the nonfat dry milk powder! Yes, you can now find fat-free sour cream just about everywhere. And, it's great on a baked potato. But, have you noticed the texture isn't quite as creamy as real sour cream? **JO's Sour Cream** has a texture closer to the real thing, so it works better in dips and baked goods.

JO's Sweetened Condensed Milk

½ cup cold water
1⅓ cups Carnation Nonfat Dry Milk Powder
½ cup Sugar Twin or Sprinkle Sweet

Place cold water in a 2-cup glass measuring cup. Stir in dry milk powder until mixture makes a smooth paste. Cover and microwave on HIGH (100% power) for 45 to 60 seconds or until mixture is very hot, but not to the boiling point. Stir in Sugar Twin. Mix well to combine. Cover and refrigerate for at least 2 hours before using. Will keep up to 2 weeks in refrigerator. Use in any recipe that calls for sweetened condensed milk. Makes equivalent of one 12-fluid-ounce can of commercial brand.

The entire recipe equals:

HE: 4 Skim Milk • 44 Optional Calories

316 Calories • 0 gm Fat • 32 gm Protein •
47 gm Carbohydrate • 496 mg Sodium •
1112 mg Calcium • 0 gm Fiber

DIABETIC: 4 Skim Milk

Just about the *most unhealthy* ingredient found in our grocery stores went to "reform school" and graduated a winner in nutrition! Fat-free sweetened condensed milk is now available in most grocery stores, but it is still so high in sugar that it's not a good choice for diabetics or people concerned about excess calories. Besides, my version is much less expensive.

JO's Cream Sauce

1 cup Carnation Nonfat Dry Milk Powder
1½ cups water
3 tablespoons all-purpose flour
1 teaspoon dried parsley flakes
¼ teaspoon black pepper

In a covered jar, combine dry milk powder, water, and flour. Cover and shake well to blend. Add parsley flakes and black pepper. Cover and shake again to blend. Pour mixture into a medium saucepan sprayed with butter-flavored cooking spray. Cook over medium heat until mixture thickens, stirring constantly. Remove from heat. Use in any recipe that requires a medium cream sauce. Makes 1½ cups.

The entire recipe equals:

HE: 3 Skim Milk • 1 Bread

329 Calories • 1 gm Fat • 26 gm Protein •
54 gm Carbohydrate • 376 mg Sodium •
848 mg Calcium • 1 gm Fiber

DIABETIC: 3 Skim Milk • 1 Starch

After preparing my cream sauce, you will never go back to making a roux! I don't believe there is an easier or healthier cream sauce anywhere in "Recipe Land."

Variations:

JO's Cheesy Cream Sauce
This cheesy cream sauce is great in traditional casseroles.
Add ¾ cup (3 ounces) shredded reduced-fat Cheddar or Swiss cheese after pouring mixture into saucepan. Continue cooking as usual.

The entire recipe equals:

HE: 4 Protein • 3 Skim Milk • 1 Bread

---

546 Calories • 14 gm Fat • 48 gm Protein •
57 gm Carbohydrate • 1089 mg Sodium •
1408 mg Calcium • 1 gm Fiber

DIABETIC: 3 Meat • 3 Skim Milk • 1 Starch

JO's Mushroom Sauce

This can be used in any recipe that calls for a can of cream of mushroom soup.

Add ½ cup (one 2.5-ounce jar) drained sliced mushrooms after pouring mixture into saucepan. Continue cooking as usual.

The entire recipe equals:

HE: 3 Skim Milk • 1 Bread • 1 Vegetable

---

353 Calories • 1 gm Fat • 28 gm Protein •
58 gm Carbohydrate • 708 mg Sodium •
857 mg Calcium • 3 gm Fiber

DIABETIC: 3 Skim Milk • 1 Starch • 1 Vegetable

JO's Peanut Sauce

This sauce is "out of this world" spooned over boiled cabbage or in any oriental dish.

Add ¼ cup reduced-fat peanut butter and 2 teaspoons reduced-sodium soy sauce after pouring mixture into saucepan. Continue cooking as usual.

The entire recipe equals:

HE: 4 Protein • 4 Fat • 3 Skim Milk • 1 Bread

---

715 Calories • 23 gm Fat • 43 gm Protein •
84 gm Carbohydrate • 1,009 mg Sodium •
848 mg Calcium • 5 gm Fiber

DIABETIC: 4 Meat • 4 Fat • 3 Skim Milk • 1 Starch

JO's Deliteful Whipped Topping

¾ cup Yoplait plain fat-free yogurt
⅓ cup Carnation Nonfat Dry Milk Powder
1 teaspoon vanilla extract
Sugar substitute to equal ¼ cup sugar
1 cup Cool Whip Free

In a medium bowl, combine yogurt and dry milk powder. Add vanilla extract and sugar substitute. Mix well to combine. Fold in Cool Whip Free. Use as you would purchased whipped topping. Makes 2 cups.

The entire recipe equals:

HE: 2 Skim Milk • 144 Optional Calories

292 Calories • 0 gm Fat • 18 gm Protein •
55 gm Carbohydrate • 291 mg Sodium •
615 mg Calcium • 0 gm Fiber

DIABETIC: 2 Skim Milk

You now have whipped topping to use in any way you would use regular whipped topping. This is close enough in volume to use for any recipe that calls for an 8-ounce container of whipped topping. Store in a covered container and use a spoonful at a time or the "whole thing" for your old-time creamy salad recipes. The addition of yogurt and dry milk powder cut the coconut and palm oils to almost nothing while increasing the calcium 200 percent. This makes a marginal nutritional choice into a high-calcium treat. The texture is almost a cross between marshmallow cream and whipped cream.

JO's Dry Casserole Soup Mix

2 cups Carnation Nonfat Dry Milk Powder

1 cup cornstarch

½ cup dry bouillon (your choice of flavor)

5 tablespoons dry onion flakes

1 tablespoon dry thyme

1 tablespoon dry basil

1 tablespoon black pepper

In a large bowl, combine dry milk powder, cornstarch, and dry bouillon. Add onion flakes, thyme, basil, and black pepper. Mix well to combine. Store in a covered container. To use, combine ⅓ cup soup mix and 1 cup water. Stir well and use as you would cream soups.

⅓ cup mix equals:

HE: ½ Skim Milk • 31 Optional Calories

---

76 Calories • 0 gm Fat • 3 gm Protein •
16 gm Carbohydrate • 466 mg Sodium •
110 mg Calcium • 0 gm Fiber

DIABETIC: ½ Skim Milk

This is a nice alternative to use if you are watching your sodium as it still gives your final product that creamy texture the purchased soups do.

JO's Milk Beverage Mixes

JO's Chocolate Milk Beverage Mix

⅓ cup Carnation Nonfat Dry Milk Powder
2 teaspoons unsweetened cocoa
Sugar substitute to equal 3 tablespoons sugar

In a small bowl, combine dry milk powder and cocoa. Add sugar substitute. Mix well to combine.

The entire recipe equals:

HE: 1 Skim Milk • 6 Optional Calories

104 Calories • 0 gm Fat • 8 gm Protein •
18 gm Carbohydrate • 133 mg Sodium •
280 mg Calcium • 1 gm Fiber

DIABETIC: 1 Skim Milk

JO's Vanilla Milk Beverage Mix

⅓ cup Carnation Nonfat Dry Milk Powder
Sugar substitute to equal 2 tablespoons sugar
1 teaspoon vanilla extract

In a small bowl, combine dry milk powder and sugar substitute. When adding liquid to mixture, stir in vanilla extract at same time.

The entire recipe equals:

HE: 1 Skim Milk • 6 Optional Calories

92 Calories • 0 gm Fat • 8 gm Protein •
15 gm Carbohydrate • 129 mg Sodium •
276 mg Calcium • 0 gm Fiber

DIABETIC: 1 Skim Milk

JO's Strawberry Milk Beverage Mix

⅓ cup Carnation Nonfat Dry Milk Powder
Sugar substitute to equal 2 tablespoons sugar
1 teaspoon strawberry extract
3 to 4 drops red food coloring

In a small bowl, combine dry milk powder and sugar substitute. When adding liquid to mixture, stir in strawberry extract and red food coloring at the same time.

The entire recipe equals:

HE: 1 Skim Milk • 6 Optional Calories

92 Calories • 0 gm Fat • 8 gm Protein •
15 gm Carbohydrate • 129 mg Sodium •
276 mg Calcium • 0 gm Fiber

DIABETIC: 1 Skim Milk

Each beverage mix equals an individual packet of purchased beverage mix.

If your original recipe calls for two packets, double everything. If your final product doesn't require heat, use an aspartame-based sugar substitute. But if you are making a brownie or anything that requires cooking or baking, use a saccharin-based sugar substitute.

To stir up your own glass of flavored milk, combine the **JO's Milk Beverage Mix** of your choice and 1 cup very cold water in a glass. Mix well.

If you want to prepare one of those wonderful shakes the original beverage mix was famous for, combine **JO's Milk Beverage Mix**, ¾ cup cold water, and 3 to 4 ice cubes in a blender container and process on BLEND until smooth.

So dig out those old recipes and toast them with a glass of **JO's Milk Beverage Mix**!

JO's Thicker Instant Pudding

1 (4-serving) package JELL-O sugar-free instant pudding mix
(any flavor)
²⁄₃ cup Carnation Nonfat Dry Milk Powder
1½ cups water
¼ cup Cool Whip Free

In a large bowl, combine dry pudding mix and dry milk powder. Add water. Mix well using a wire whisk. Blend in Cool Whip Free. Use in any recipe that calls for a prepared package of instant pudding.

The entire recipe equals:

HE: 2 Skim Milk • 140 Optional Calories

284 Calories • 0 gm Fat • 16 gm Protein •
55 gm Carbohydrate • 1576 mg Sodium •
552 mg Calcium • 0 gm Fiber

DIABETIC: 2 Skim Milk • 1½ Starch

My version of instant pudding gives a much creamier taste and is guaranteed not to separate when made into a pie filling, the way instant pudding usually does when prepared with skim milk. This also sets up quicker, usually in 5 minutes or less. So, if someone is unexpectedly knocking at your door at mealtime, you can quickly throw a delicious pie together and enjoy it only minutes later!

JO's Thicker Cooked Pudding

1 (4-serving) package JELL-O sugar-free cook-and-serve
    pudding mix (any flavor)
⅔ cup Carnation Nonfat Dry Milk Powder
1½ cups water

In a medium saucepan, combine dry pudding mix and
dry milk powder. Add water. Cook over medium heat until mixture thickens and starts to boil, stirring constantly. Remove from
heat. Use in any recipe that calls for a prepared package of cooked
pudding.

The entire recipe equals:

HE: 2 Skim Milk • 80 Optional Calories

236 Calories • 0 gm Fat • 16 gm Protein •
43 gm Carbohydrate • 706 gm Sodium •
552 mg Calcium • 0 gm Fiber

DIABETIC: 2 Skim Milk • 1 Starch

My version of cooked pudding is so soft and velvety, you will never
believe cream isn't one of the main ingredients.

Soups

"Cream" of Tomato Soup

This is so "um-um-good," it'll make you feel all cozy and satisfied, just as Mom's soup-and-sandwich lunches always made you feel loved. For those of you who prefer your tomato soup more tangy than sweet, you may want to try adding just one teaspoon of Sugar Twin. ☺ Serves 2 (1¼ cups)

> 1¾ cups (one 15-ounce can) Hunt's Chunky Tomato Sauce
> ½ teaspoon baking soda
> 2 teaspoons Sugar Twin or Sprinkle Sweet
> ⅛ teaspoon black pepper
> ½ teaspoon Italian seasoning
> **2 recipes JO's Evaporated Skim Milk**

In a medium saucepan, combine tomato sauce and baking soda. Bring mixture to a boil. Stir in Sugar Twin, black pepper, and Italian seasoning. Add **JO's Evaporated Skim Milk**. Mix well to combine. Lower heat and simmer for 4 to 5 minutes or until mixture is heated through, stirring occasionally.

Each serving equals:

HE: 3½ Vegetable • 1 Skim Milk • 2 Optional Calories

128 Calories • 0 gm Fat • 10 gm Protein •
22 gm Carbohydrate • 938 mg Sodium •
276 mg Calcium • 3 gm Fiber

DIABETIC: 2 Vegetable • ½ Skim Milk

Quick Carrot-Potato Soup

Your kids might want to call this luscious concoction "Mashed Potato" Soup, but whatever its name, it's got lots of creamy flavor that's bound to make them—and you—smile!

● Serves 4 (1½ cups)

2 cups (one 16-ounce can) Healthy Request Chicken Broth

¾ cup chopped celery

¾ cup chopped onion

2 cups diced fresh carrots

1 (8-ounce) package Philadelphia fat-free cream cheese

1½ cups (one 12-fluid-ounce can) Carnation Evaporated Skim Milk

⅔ cup (1½ ounces) instant potato flakes

1 teaspoon dried parsley flakes

¼ teaspoon black pepper

In a large saucepan, combine chicken broth, celery, onion, and carrots. Bring mixture to a boil. Lower heat, cover and simmer for 15 minutes or until vegetables are tender. Stir in cream cheese and evaporated skim milk. Add potato flakes, parsley flakes, and black pepper. Mix well to combine. Continue simmering for 5 to 6 minutes or until cream cheese melts and mixture thickens, stirring often.

Each serving equals:

HE: 1¾ Vegetable • 1 Protein • ¾ Skim Milk • ½ Bread • 8 Optional Calories

---

184 Calories • 0 gm Fat • 18 gm Protein • 28 gm Carbohydrate • 638 mg Sodium • 312 mg Calcium • 3 gm Fiber

---

DIABETIC: 1 Vegetable • 1 Meat • 1 Skim Milk • ½ Starch

Homestyle Potato Soup

Here's a rich potato soup that's even more scrumptious with the addition of corn and a touch of dill and lemon pepper for a unique flavor. If you rarely add any spices besides salt and pepper, you'll be delighted to discover how just a pinch of seasoning delivers great taste! ☻ Serves 8 (1 cup)

2½ cups water
4 cups (two 16-ounce cans) Healthy Request Chicken Broth☆
1 cup chopped onion
2 cups chopped celery
4 cups (20 ounces) diced raw potatoes
1 cup frozen whole kernel corn
1 tablespoon dried parsley flakes
1 teaspoon dried dill weed
1⅓ cups Carnation Nonfat Dry Milk Powder
6 tablespoons all-purpose flour
1 tablespoon Sugar Twin or Sprinkle Sweet
½ teaspoon lemon pepper

In a large saucepan, combine water, 2 cups chicken broth, onion, celery, potatoes, and corn. Bring mixture to a boil. Stir in parsley flakes and dill weed. Lower heat, cover, and simmer for 20 minutes or until vegetables are tender. In a medium bowl, combine remaining 2 cups chicken broth, dry milk powder, flour, Sugar Twin, and lemon pepper. Mix well using a wire whisk. Stir broth mixture into potato mixture. Continue cooking for 5 to 6 minutes or until mixture thickens, stirring often.

Each serving equals:

HE: 1 Bread • ¾ Vegetable • ½ Skim Milk •
9 Optional Calories

140 Calories • 0 gm Fat • 8 gm Protein •
27 gm Carbohydrate • 334 mg Sodium •
166 mg Calcium • 2 gm Fiber

DIABETIC: 1½ Starch • ½ Skim Milk

Tomato Neapolitan Corn Chowder

Did you know that "Neapolitan" is the nickname for all the delectable dishes created for us by the chefs of Naples, Italy? This hearty blend of tastes will make you feel like bursting into song and dance. Tarantella, anyone? ☻ Serves 4 (1½ cups)

2 cups (10 ounces) diced raw potatoes
2 cups (one 16-ounce can) Healthy Request Chicken Broth
1 cup frozen whole kernel corn
1 (10 ¾-ounce) can Healthy Request Tomato Soup
½ cup (one 2.5-ounce jar) sliced mushrooms, drained
1 teaspoon Italian seasoning
2 teaspoons Sugar Twin or Sprinkle Sweet
½ teaspoon dried minced garlic
2 teaspoons dried parsley flakes
2 recipes JO's Evaporated Skim Milk

In a large saucepan, combine potatoes and chicken broth. Bring mixture to a boil. Lower heat, cover and simmer for 15 minutes or until potatoes are tender. Add corn, tomato soup, mushrooms, Italian seasoning, Sugar Twin, garlic, and parsley flakes. Mix well to combine. Stir **JO's Evaporated Skim Milk** into corn mixture. Continue cooking for 10 minutes or until mixture is heated through, stirring occasionally.

Each serving equals:

HE: 1 Bread • ½ Skim Milk • ¼ Vegetable •
¼ Slider • 14 Optional Calories

181 Calories • 1 gm Fat • 9 gm Protein •
34 gm Carbohydrate • 621 mg Sodium •
157 mg Calcium • 3 gm Fiber

DIABETIC: 1½ Starch • ½ Skim Milk • ½ Vegetable

Milwaukee Cheese-Veggie Soup ❄

Doesn't the thought of cheese soup on a cool fall day just warm you through and through? This dish is so cheesy, thick, and filling, no one will believe how healthy it is, too! Go on, chase the chill away with this steamy soup! ☻ Serves 4 (1 cup)

2½ cups frozen carrot, broccoli, and cauliflower blend
1½ cups water☆
2 teaspoons dried onion flakes
2 teaspoons dried parsley flakes
1 (10 ¾-ounce) can Healthy Request Cream of Mushroom Soup
⅔ cup Carnation Nonfat Dry Milk Powder
¼ cup Lite Beer or water
1½ cups (6 ounces) shredded Kraft reduced-fat Cheddar cheese

In a medium saucepan, combine frozen vegetables and ½ cup water. Cook over medium heat for 6 to 8 minutes or until vegetables are just tender. Stir in onion flakes and parsley flakes. Add mushroom soup, dry milk powder, remaining 1 cup water, and beer. Mix well to combine. Stir in Cheddar cheese. Lower heat and simmer for 5 to 6 minutes or until cheese melts, stirring often.

HINT: ½ cup frozen carrots, 1 cup frozen broccoli, and 1 cup frozen cauliflower may be used in place of blended vegetables.

Each serving equals:

HE: 2 Protein • 1¼ Vegetable • ½ Skim Milk •
½ Slider • 7 Optional Calories

216 Calories • 8 gm Fat • 17 gm Protein •
19 gm Carbohydrate • 637 mg Sodium •
503 mg Calcium • 2 gm Fiber

DIABETIC: 1½ Meat • 1 Vegetable • 1 Starch •
½ Skim Milk

Creamy Italian Tomato Rice Soup

This tastes amazingly fresh, but you don't need to wait for the abundance of a late-summer tomato harvest! The canned ones cook up tangy and fast, and your bones will thank you for serving this often.

○ Serves 4 (1¼ cups)

½ cup finely chopped onion
½ cup finely chopped green bell pepper
4 cups (two 16-ounce cans) chopped tomatoes, undrained
2 teaspoons Italian seasoning
⅛ teaspoon black pepper
2 recipes JO's Evaporated Skim Milk
⅔ cup (2 ounces) uncooked instant rice

In a large saucepan sprayed with olive oil–flavored cooking spray, sauté onion and green pepper for 5 minutes or until tender. Meanwhile, place undrained tomatoes in a blender container. Cover and process on HIGH for 15 seconds or until pureed. Add pureed tomatoes to onion mixture. Mix well to combine. Stir in Italian seasoning and black pepper. Cook over medium heat until mixture comes to a boil, stirring often. Lower heat and simmer 5 minutes. Add **JO's Evaporated Skim Milk** and rice to tomato mixture. Mix gently to combine. Continue simmering for 6 to 8 minutes or until rice is tender, stirring occasionally.

HINT: If you like tomatoes on the sweeter side, add 1 tablespoon Sugar Twin when adding rice.

Each serving equals:

HE: 2½ Vegetable • ½ Bread • ½ Skim Milk

---

121 Calories • 1 gm Fat • 6 gm Protein •
22 gm Carbohydrate • 79 mg Sodium •
155 mg Calcium • 2 gm Fiber

DIABETIC: 2 Vegetable • 1 Starch • ½ Skim Milk

Broccoli Farmstead Chowder

This soup just overflows with Midwestern goodness, a blend of harvest flavors you can enjoy anytime. Becky loves broccoli—and this chowder—as much as I do. Like mother, like daughter!

Serves 6 (1½ cups)

2 cups (one 16-ounce can) Healthy Request Chicken Broth
3 cups frozen chopped broccoli
¾ cup chopped onion
1 full cup (6 ounces) diced cooked chicken breast
2 cups (one 16-ounce can) cream-style corn
1 cup Carnation Nonfat Dry Milk Powder
⅔ cup (1½ ounces) instant potato flakes
1 teaspoon dried parsley flakes
⅛ teaspoon black pepper
1 cup water

In a large saucepan, combine chicken broth, broccoli, and onion. Bring mixture to a boil. Lower heat and simmer for 10 minutes or until broccoli is tender. DO NOT DRAIN. Stir in chicken and corn. In a small bowl, combine dry milk powder, potato flakes, parsley flakes, black pepper, and water. Add milk mixture to broccoli mixture. Mix well to combine. Continue simmering for 10 minutes or until mixture thickens and is heated through, stirring occasionally.

Each serving equals:

HE: 1¼ Vegetable • 1 Protein • 1 Bread •
½ Skim Milk • 5 Optional Calories

197 Calories • 1 gm Fat • 17 gm Protein •
30 gm Carbohydrate • 505 mg Sodium •
174 mg Calcium • 3 gm Fiber

DIABETIC: 1½ Starch • 1 Meat • 1 Vegetable •
½ Skim Milk

Calico Chicken Vegetable Soup

I always like to "layer" the flavors in my recipes, so you get the tastiest dishes possible. This is a good example, as it includes chunks of chicken, chicken broth, cream of chicken soup, and poultry seasoning. Yummy! ☻ Serves 4 (1½ cups)

2 cups (one 16-ounce can) Healthy Request Chicken Broth

1 full cup (6 ounces) diced cooked chicken breast

3 cups frozen carrot, broccoli, and cauliflower blend

½ teaspoon dried poultry seasoning

1 teaspoon dried parsley flakes

¼ teaspoon black pepper

1 (10¾-ounce) can Healthy Request Cream of Chicken Soup

2 recipes JO's Evaporated Skim Milk

⅓ cup (1 ounce) uncooked instant rice

In a large saucepan, combine chicken broth, chicken, frozen vegetables, poultry seasoning, parsley flakes, and black pepper. Bring mixture to a boil. Lower heat, cover, and simmer 15 minutes or until vegetables are tender. Meanwhile, in a small bowl, combine chicken soup and **JO's Evaporated Skim Milk**. Add soup mixture to vegetable mixture. Mix well to combine. Continue cooking for 10 minutes, stirring occasionally. Stir in rice. Re-cover, remove from heat, and let set for 5 minutes. Mix well before serving.

HINT: 1 cup frozen carrots, 1 cup frozen broccoli, and 1 cup frozen cauliflower or any other combination of frozen vegetables may be used.

Each serving equals:

HE: 1½ Protein • 1½ Vegetable • ½ Skim Milk •
¼ Bread • ½ Slider • 13 Optional Calories

207 Calories • 3 gm Fat • 22 gm Protein •
23 gm Carbohydrate • 657 mg Sodium •
169 mg Calcium • 2 gm Fiber

DIABETIC: 1½ Meat • ½ Vegetable • ½ Starch •
½ Skim Milk

Corn Ham Chowder

I know what you're thinking—how can a recipe that includes cream soup, creamy corn, and evaporated milk be so good for you? Well, with the help of my magic whisk, I've whisked out the fat—but kept all those scrumptious flavors just for you.

○ Serves 4 (1½ cups)

> 1 cup water
> 1 cup (5 ounces) diced raw potatoes
> ½ cup chopped onion
> 1 full cup (6 ounces) diced Dubuque 97% fat-free ham or
>   any extra-lean ham
> 2 cups (one 16-ounce can) cream-style corn
> 1½ cups (one 12-fluid-ounce can) Carnation Evaporated
>   Skim Milk
> 1 (10 ¾-ounce) can Healthy Request Cream of
>   Mushroom Soup
> 1 teaspoon dried parsley flakes
> ¼ teaspoon black pepper

In a large saucepan, combine water, potatoes, and onion. Bring mixture to a boil. Lower heat, cover, and simmer for 10 minutes or until potatoes are just tender. Stir in ham and corn. Add evaporated skim milk, mushroom soup, parsley flakes, and black pepper. Mix well to combine. Continue simmering for 15 minutes or until mixture is heated through, stirring occasionally.

Each serving equals:

HE: 1¼ Bread • 1 Protein • ¾ Skim Milk •
¼ Vegetable • ½ Slider • 1 Optional Calorie

288 Calories • 4 gm Fat • 18 gm Protein •
45 gm Carbohydrate • 938 mg Sodium •
340 mg Calcium • 2 gm Fiber

DIABETIC: 2 Starch • 1 Meat • 1 Skim Milk

Tom's Green Bean Chowder ❄

My husband, Cliff, and son Tommy both love green beans, so I'm always looking for new ways to use them in my recipes. This is a real man-pleaser, blending corn and potatoes to add heartiness and lip-smacking fun! ◐ Serves 4 (1½ cups)

> 8 ounces ground 90% lean turkey or beef
> 2 cups (one 16-ounce can) cut green beans,
>   rinsed and drained
> ½ cup (one 2.5-ounce jar) sliced mushrooms, drained
> 2 cups (one 16-ounce can) cream-style corn
> 2 cups skim milk
> 1 teaspoon dried parsley flakes
> ⅛ teaspoon black pepper
> ⅓ cup (¾ ounce) instant potato flakes

In a large saucepan sprayed with butter-flavored cooking spray, brown meat. Add green beans, mushrooms, corn, skim milk, parsley flakes, and black pepper. Mix well to combine. Stir in potato flakes. Lower heat and simmer for 15 minutes, stirring occasionally.

Each serving equals:

HE: 1½ Protein • 1¼ Vegetable • 1¼ Bread •
½ Skim Milk

---

261 Calories • 5 gm Fat • 18 gm Protein •
36 gm Carbohydrate • 570 gm Sodium •
178 mg Calcium • 3 gm Fiber

DIABETIC: 1½ Meat • 1½ Starch • 1 Vegetable •
½ Skim Milk

Cold Cucumber Soup

So quick, so cool, and oh-so-creamy—here's a pretty-to-look-at and tasty way to enjoy the healthy benefits of your garden. If you haven't got an extra couple of cukes, you know every farmstand is overflowing with them at irresistible prices!

◐ Serves 2 (1 full cup)

½ teaspoon minced garlic
1½ teaspoons dried dill weed
⅓ cup Carnation Nonfat Dry Milk Powder
¾ cup Yoplait plain fat-free yogurt
1½ cups unpeeled chopped cucumbers

In a blender container, combine garlic, dill weed, dry milk powder, yogurt, and cucumbers. Cover and process on BLEND for 30 seconds or until mixture is smooth. Pour mixture into 2 soup bowls. Cover and refrigerate for at least 15 minutes.

HINT: Good served garnished with 1 tablespoon fat-free sour cream, but don't forget to count the few additional calories.

Each serving equals:

HE: 1½ Vegetable • 1 Skim Milk

92 Calories • 0 gm Fat • 10 gm Protein •
13 gm Carbohydrate • 130 mg Sodium •
342 mg Calcium • 1 gm Fiber

DIABETIC: 1 Vegetable • 1 Skim Milk

Chilled Strawberry Soup

Anyone who knows me knows I could eat strawberries all day long! This is a great summer soup that makes a lovely dessert, too—or a delectably inventive addition to a bridal shower menu.

○ Serves 6 (1 cup)

3 cups frozen unsweetened strawberries, thawed
1 cup skim milk
¾ cup Yoplait plain fat-free yogurt
¼ cup Cary's Sugar Free Maple Syrup
½ teaspoon vanilla extract
1 (4-serving) package JELL-O sugar-free strawberry gelatin
2 tablespoons (½ ounce) chopped pecans

Reserve ½ cup strawberries. In a blender container, combine remaining strawberries with juice and skim milk. Cover and process on HIGH for 20 seconds or until mixture is smooth. Pour mixture into a large bowl. Add yogurt, maple syrup, and vanilla extract. Mix well using a wire whisk. Stir in dry gelatin. Coarsely chop remaining ½ cup strawberries. Fold chopped strawberries into soup mixture. Refrigerate for at least 30 minutes. When serving, spoon mixture into soup bowls and top each with 1 teaspoon pecans.

HINT: Also good garnished with 1 tablespoon Cool Whip Lite, but don't forget to count the few additional calories.

Each serving equals:

HE: ½ Fruit • ⅓ Skim Milk • ⅓ Fat •
13 Optional Calories

82 Calories • 2 gm Fat • 4 gm Protein •
12 gm Carbohydrates • 102 mg Sodium •
118 mg Calcium • 1 gm Fiber

DIABETIC: ½ Fruit • ½ Skim Milk • ½ Fat

Vegetables

Asparagus and Mushroom Casserole

Evaporated skim milk is one of my favorite ways to add calcium and creamy goodness to a tasty vegetable casserole, and Becky agreed that this one is elegant, too. It's crunchy and tangy, full of good nutrition, and easy on your calorie budget.

● Serves 4

> 1½ cups (one 12-fluid-ounce can) Carnation Evaporated
>   Skim Milk
> 3 tablespoons all-purpose flour
> 1 teaspoon dried onion flakes
> ¼ teaspoon black pepper
> 1¾ cups (one 14.5-ounce can) cut asparagus,
>   rinsed and drained
> ¾ cup chopped fresh mushrooms
> 14 small fat-free saltine crackers, coarsely crushed
> 2 hard-boiled eggs, diced
> 2 tablespoons (½ ounce) chopped pecans

Preheat oven to 350 degrees. Spray an 8-by-8-inch baking dish with butter-flavored cooking spray. In a covered jar, combine evaporated skim milk and flour. Shake well to blend. Pour mixture into a medium saucepan sprayed with butter-flavored cooking spray. Cook over medium heat for 5 minutes or until mixture thickens, stirring often. Stir in onion flakes and black pepper. Remove from heat. Layer asparagus, fresh mushrooms, crackers, and eggs in prepared baking dish. Spoon cream sauce over top. Evenly sprinkle pecans over sauce. Bake for 30 minutes. Place baking dish on a wire rack and let set for 5 minutes. Divide into 4 servings.

Each serving equals:

HE: 1¼ Vegetable • ¾ Skim Milk • ¾ Bread •
½ Fat • ½ Protein (limited)

260 Calories • 8 gm Fat • 15 gm Protein •
32 gm Carbohydrate • 403 mg Sodium •
334 mg Calcium • 4 gm Fiber

DIABETIC: 1½ Vegetable • 1 Skim Milk • 1 Starch •
½ Fat • ½ Meat

Green Beans Extraordinaire

If you've ever ordered a dish prepared "Almondine," you know the French were definitely on to something! Green beans are a regular on our menu, but here the almonds give this cozy home-style green bean dish a little "extra" sparkle! ❂ Serves 6 (1 cup)

> 5 cups frozen cut green beans
> 3¼ cups water☆
> ½ cup chopped onion
> 1½ cups sliced fresh mushrooms
> ¼ cup (1 ounce) slivered almonds
> ¼ teaspoon black pepper
> 1 (10¾ ounce) can Healthy Request Cream of
>   Mushroom Soup
> ⅔ cup Carnation Nonfat Dry Milk Powder

In a large saucepan, cook green beans in 2½ cups water for 10 minutes or just until tender. Drain. Meanwhile, in a large skillet sprayed with butter-flavored cooking spray, sauté onion for 5 minutes. Stir in mushrooms and almonds. Continue sautéing for 3 to 4 minutes. Add drained green beans and black pepper. Mix well to combine. In a small bowl, combine mushroom soup, dry milk powder, and remaining ¾ cup water. Stir soup mixture into green bean mixture. Lower heat and simmer for 3 to 4 minutes, or until mixture is heated through, stirring often.

Each serving equals:

HE: 2⅓ Vegetable • ⅓ Fat • ⅓ Skim Milk •
¼ Slider • 19 Optional Calories

126 Calories • 4 gm Fat • 6 gm Protein •
19 gm Carbohydrate • 412 mg Sodium •
174 mg Calcium • 4 gm Fiber

DIABETIC: 2 Vegetable • ½ Fat • ½ Starch

Green Beans Fiesta

Here's a wonderful side dish to add South-of-the-Border pizazz to any meal. Your taste buds will quickly shout "Olé" after just one bite!　　❂　Serves 4

> 1 (10¾-ounce) can Healthy Request Cream of
> 　　Mushroom Soup
> ⅓ cup Carnation Nonfat Dry Milk Powder
> ½ cup chunky salsa (mild, medium, or hot)
> 4 cups (two 16-ounce cans) cut green beans,
> 　　rinsed and drained
> ½ cup (one 2.5-ounce jar) sliced mushrooms, drained
> 6 tablespoons (1½ ounces) dried fine bread crumbs
> ½ teaspoon chili seasoning
> 1 teaspoon dried parsley flakes

Preheat oven to 350 degrees. Spray an 8-by-8-inch baking dish with butter-flavored cooking spray. In a large bowl, combine mushroom soup, dry milk powder, and salsa. Add green beans and mushrooms. Mix well to combine. Pour mixture into prepared baking dish. In a small bowl, combine bread crumbs, chili seasoning, and parsley flakes. Sprinkle bread mixture evenly over top. Lightly spray crumbs with butter-flavored cooking spray. Bake for 30 minutes. Place baking dish on a wire rack and let set for 5 minutes. Divide into 4 servings.

Each serving equals:

HE: 2½ Vegetable • ½ Bread • ¼ Skim Milk •
½ Slider • 1 Optional Calorie

---

142 Calories • 2 gm Fat • 6 gm Protein •
25 gm Carbohydrate • 618 mg Sodium •
220 mg Calcium • 3 gm Fiber

DIABETIC: 2½ Vegetable • 1 Starch

Escalloped Cabbage

Cheesy, creamy, crunchy—a party of flavors in one easy-to-fix delight! And just look how much calcium I've stirred into this delicious side dish.   ☻   Serves 4

    3½ cups shredded cabbage
    ½ cup chopped onion
    1½ cups hot water
    1½ cups (one 12-fluid-ounce can) Carnation Evaporated
        Skim Milk
    3 tablespoons all-purpose flour
    ½ teaspoon lemon pepper
    ¾ cup (3 ounces) shredded Kraft reduced-fat Cheddar cheese
    14 fat-free saltine crackers, crumbled

Preheat oven to 350 degrees. Spray an 8-by-8-inch baking dish with butter-flavored cooking spray. In a medium saucepan, cook cabbage and onion in water for 5 minutes. Meanwhile, in a covered jar, combine evaporated skim milk and flour. Shake well to blend. Pour milk mixture into a large saucepan sprayed with butter-flavored cooking spray. Stir in lemon pepper. Cook over medium heat for 5 minutes or until mixture thickens, stirring often. Drain cabbage mixture and add to cream sauce. Mix gently to combine. Place half of mixture in prepared baking dish. Sprinkle half of Cheddar cheese evenly over top. Repeat layers. Evenly sprinkle cracker crumbs over top. Lightly spray crumbs with butter-flavored cooking spray. Bake for 30 minutes. Place baking dish on a wire rack and let set for 5 minutes. Divide into 4 servings.

Each serving equals:

HE: 2 Vegetable • 1 Protein • ¾ Bread • ¾ Skim Milk

207 Calories • 3 gm Fat • 16 gm Protein •
29 gm Carbohydrate • 424 mg Sodium •
448 mg Calcium • 2 gm Fiber

DIABETIC: 1 Vegetable • 1 Protein • 1 Starch •
1 Skim Milk

Carrot and Celery au Gratin

What could be more irresistible than your favorite crunchy carrots and celery blended into a warm and cozy old-fashioned hot dish? This one bubbles with cheesy goodness and fills the house with a scrumptious aroma.   ☻   Serves 6

3 cups sliced carrots
2 cups sliced celery
2 cups hot water
1 (10¾-ounce) can Healthy Request Cream of Celery Soup
¾ cup (3 ounces) shredded Kraft reduced-fat Cheddar cheese
6 tablespoons (1½ ounces) dried fine bread crumbs☆

Preheat oven to 350 degrees. Spray an 8-by-8-inch baking dish with butter-flavored cooking spray. In a large saucepan, cook carrots and celery in water for 10 minutes or until vegetables are just tender. Drain. Return vegetables to saucepan. Add celery soup, Cheddar cheese, and 3 tablespoons bread crumbs. Mix well to combine. Pour mixture into prepared baking dish. Evenly sprinkle remaining bread crumbs over top. Lightly spray crumbs with butter-flavored cooking spray. Bake for 45 minutes. Place baking dish on a wire rack and let set for 5 minutes. Divide into 6 servings.

Each serving equals:

HE: 1⅔ Vegetable • ⅔ Protein • ⅓ Bread • ¼ Slider • 8 Optional Calories

120 Calories • 3 gm Fat • 6 gm Protein •
17 gm Carbohydrate • 436 mg Sodium •
174 mg Calcium • 2 gm Fiber

DIABETIC: 2 Vegetable • ½ Meat • ½ Starch

Scalloped Corn and Carrots ❇

My eternal thanks to the folks who invented a way to make saltine crackers fat-free! They're such a simple way to add texture and taste to an old-fashioned family recipe like this one. This would have been a hit at Grandma's boardinghouse. ☺ Serves 6

2 cups (one 16-ounce can) cream-style corn
½ cup chopped onion
1 cup grated carrots
¼ teaspoon black pepper
14 small fat-free saltine crackers, crushed
⅔ cup Carnation Nonfat Dry Milk Powder
½ cup water
1 teaspoon dried parsley flakes

Preheat oven to 350 degrees. Spray an 8-by-8-inch baking dish with butter-flavored cooking spray. In a medium bowl, combine corn, onion, and carrots. Stir in black pepper and cracker crumbs. In a small bowl, combine dry milk powder, water, and parsley flakes. Add milk mixture to corn mixture. Mix well to combine. Pour mixture into prepared baking dish. Bake for 45 minutes. Place baking dish on a wire rack and let set 5 minutes. Divide into 6 servings.

Each serving equals:

HE: 1 Bread • ½ Vegetable • ⅓ Skim Milk

132 Calories • 0 gm Fat • 5 gm Protein •
28 gm Carbohydrate • 345 mg Sodium •
118 mg Calcium • 2 gm Fiber

DIABETIC: 2 Starch

Corn Pudding

Here's a veggie dish that "eats" like a hearty entree! It cooks up golden brown, and I bet it'll call your family to dinner as easily as the loudest dinner bell! ☻ Serves 6

2 eggs or equivalent in egg substitute

⅔ cup Carnation Nonfat Dry Milk Powder

¾ cup water

3 cups frozen whole kernel corn, thawed

¼ cup chopped green bell pepper

½ cup chopped onion

1 teaspoon dried parsley flakes

¼ teaspoon black pepper

1 teaspoon prepared mustard

1 teaspoon Sugar Twin or Sprinkle Sweet

Preheat oven to 350 degrees. Spray an 8-by-8-inch baking dish with butter-flavored cooking spray. In a large bowl, combine eggs, dry milk powder, and water. Add corn, green pepper, onion, parsley flakes, black pepper, mustard, and Sugar Twin. Mix well to combine. Pour mixture into prepared baking dish. Bake 1 hour or until firm in center. Place baking dish on a wire rack and let set for 5 minutes. Divide into 6 servings.

HINT: Thaw corn by placing in a colander and rinsing under hot water for one minute.

Each serving equals:

HE: 1 Bread • ⅓ Protein (limited) • ⅓ Skim Milk • ¼ Vegetable

134 Calories • 2 gm Fat • 7 gm Protein • 22 gm Carbohydrate • 78 mg Sodium • 107 mg Calcium • 2 gm Fiber

DIABETIC: 1½ Starch • ½ Meat

Creamy Rice and Peas

The peas provide a pretty touch of springtime in this luscious dish that was inspired just a bit by the Italian specialty called risotto. It's a fantastic recipe to call on when you've only got ten minutes to make supper on a busy weeknight. ❂ Serves 4 (¾ cup)

½ cup (one 2.5-ounce jar) sliced mushrooms, drained
1 (10¾-ounce) can Healthy Request Cream of
   Mushroom Soup
2 recipes JO's Evaporated Skim Milk
¼ teaspoon black pepper
½ cup frozen peas, thawed
1 cup (3 ounce) uncooked instant rice
1 teaspoon dried parsley flakes

In a large skillet sprayed with butter-flavored cooking spray, combine mushrooms, mushroom soup, **JO's Evaporated Skim Milk**, and black pepper. Stir in peas. Bring mixture to a boil. Add rice and parsley flakes. Mix well to combine. Cover, remove from heat, and let set for 5 minutes. Fluff with a fork before serving.

HINT: Thaw peas by placing in a colander and rinsing under hot water for one minute.

Each serving equals:

HE: 1 Bread • ½ Skim Milk • ¼ Vegetable •
½ Slider • 1 Optional Calorie

146 Calories • 2 gm Fat • 7 gm Protein •
25 gm Carbohydrate • 447 mg Sodium •
200 mg Calcium • 2 gm Fiber

DIABETIC: 1½ Starch • ½ Skim Milk

Garden Party Potatoes

I've got a great idea—instead of going out for baked potatoes topped with tangy cheese and veggies, why not please your family anytime at all by making these special treats at home? There's almost nothing more filling or full of good nutrition than a baked potato!   ☻   Serves 6

1 recipe JO's Sour Cream
¼ cup Kraft fat-free mayonnaise
¼ cup Kraft Fat Free Ranch Dressing
1¼ cups peeled and chopped fresh tomatoes
½ cup chopped green bell pepper
¼ cup chopped onion
*2 tablespoons chopped fresh parsley or 2 teaspoons
        dried parsley flakes*
*¾ cup (1½ ounces) shredded Kraft reduced-fat
        Cheddar cheese*
6 medium-sized (5-ounce) baking potatoes

In a medium bowl, combine **JO's Sour Cream**, mayonnaise, and Ranch dressing. Stir in tomatoes, green pepper, onion, and parsley. Add Cheddar cheese. Mix well to combine. Cover and refrigerate. Meanwhile, bake potatoes in oven or microwave. When serving, split potatoes and top each with about ⅔ cup vegetable mixture.

Each serving equals:

HE: 1 Bread • ⅔ Vegetable • ⅔ Protein •
⅓ Skim Milk • ¼ Slider • 3 Optional Calories

205 Calories • 1 gm Fat • 8 gm Protein •
41 gm Carbohydrate • 308 mg Sodium •
166 mg Calcium • 3 gm Fiber

DIABETIC: 2 Starch • ½ Meat

Baked Potatoes with Bacon Topping

Try this speedy and tasty potato topper (inspired by my son James, who loves spooning toppings onto his potatoes). This bacon-chili combination is a wonderfully spicy wake-up call for your mouth.

❤ Serves 4

1 recipe JO's Sour Cream
2 teaspoons dried parsley flakes
2 tablespoons chili sauce
¼ teaspoon lemon pepper
2 tablespoons Hormel Bacon Bits
4 medium-sized (5-ounce) baking potatoes

In a medium bowl, combine **JO's Sour Cream**, parsley flakes, chili sauce, and lemon pepper. Add bacon bits. Mix gently to combine. Cover and refrigerate. Meanwhile, bake potatoes in oven or microwave. When serving, split potatoes and top each with about ¼ cup topping mixture.

Each serving equals:

HE: 1 Bread • ½ Skim Milk • ¼ Slider • 16 Optional Calories

176 Calories • 0 gm Fat • 8 gm Protein • 36 gm Carbohydrate • 183 mg Sodium • 164 mg Calcium • 3 gm Fiber

DIABETIC: 1½ Starch • ½ Skim Milk

Savory Salads

Creamy Carrot Salad

I love carrot salad for its sweetness, and also for all that healthy vitamin A! This is a classic recipe made healthy, and it belongs on every family reunion buffet table—don't you agree? This new version is extra-creamy—and low in sugar, too!

☻ Serves 4 (full ⅔ cup)

> *1 recipe JO's Sour Cream*
> *2 tablespoons Kraft fat-free mayonnaise*
> *1 teaspoon lemon juice*
> *Sugar substitute to equal 1 tablespoon sugar*
> *3 cups shredded carrots*
> *½ cup raisins*
> *2 tablespoons (½ ounce) chopped pecans*

In a large bowl, combine **JO's Sour Cream**, mayonnaise, lemon juice, and sugar substitute. Mix well to combine. Stir in carrots, raisins, and pecans. Cover and refrigerate for at least 30 minutes. Gently stir again just before serving.

HINT: To plump up raisins without "cooking," place in a glass measuring cup and microwave on HIGH for 20 seconds.

Each serving equals:

HE: 1½ Vegetable • 1 Fruit • ½ Fat • ½ Skim Milk • 3 Optional Calories

120 Calories • 1 gm Fat • 4 gm Protein • 23 gm Carbohydrate • 78 mg Sodium • 119 mg Calcium • 1 gm Fiber

DIABETIC: 1 Vegetable • 1 Fruit • ½ Fat • ½ Skim Milk

Paradise Carrot Salad

If a taste of something heavenly could actually transport you to the tropics, you'd wake up in Maui after just a few bites of this one! When the wind is blowing hard and the snow's two feet deep, why not stir up some dining "aloha" in your kitchen?

○ Serves 6 (full ¾ cup)

4 cups shredded carrots
1 cup (one 8-ounce can) crushed pineapple, packed in
*   fruit juice, drained, and 2 tablespoons liquid reserved*
½ cup (3 ounces) chopped dried apricots
1 recipe JO's Sour Cream
1 teaspoon coconut extract
Sugar substitute to equal ¼ cup sugar
½ cup Cool Whip Free

In a medium bowl, combine carrots, pineapple, and apricots. In a small bowl, combine **JO's Sour Cream**, reserved pineapple juice, coconut extract, and sugar substitute. Mix well to combine. Blend in Cool Whip Free. Add sour cream mixture to carrot mixture. Mix gently to combine. Cover and refrigerate for at least 30 minutes. Gently stir again just before serving.

HINT: To plump up apricots without "cooking," place in a glass measuring cup and microwave on HIGH for 30 seconds.

Each serving equals:

HE: 1⅓ Vegetable • 1 Fruit • ⅓ Skim Milk •
14 Optional Calories

---

136 Calories • 0 gm Fat • 4 gm Protein •
30 gm Carbohydrate • 73 mg Sodium •
134 mg Calcium • 2 gm Fiber

DIABETIC: 1½ Fruit • 1 Vegetable

Crunchy Corn-Cucumber Salad

I'm a big fan of foods that please the eye as well as the palate. This flavorful salad is richly colorful and just about perfect for a Sunday summer barbecue. ☻ Serves 6 (¾ cup)

> 2 cups frozen whole kernel corn, thawed
> 2¾ cups unpeeled thinly sliced cucumber
> ¼ cup finely chopped onion
> ¾ cup (3 ounces) shredded Kraft reduced-fat Cheddar cheese
> ½ cup Kraft fat-free mayonnaise
> 2 tablespoons white vinegar
> Sugar substitute to equal 1 tablespoon sugar
> ¼ teaspoon black pepper
> 1 tablespoon chopped fresh parsley or 1 teaspoon
>   dried parsley flakes

In a large bowl, combine corn, cucumber, onion, and Cheddar cheese. In a small bowl, combine mayonnaise, vinegar, sugar substitute, black pepper, and parsley. Add mayonnaise mixture to corn mixture. Mix gently to combine. Cover and refrigerate for at least 30 minutes. Gently stir again just before serving.

HINT: Thaw corn by placing in a colander and rinsing under hot water for one minute.

Each serving equals:

HE: 1 Vegetable • ⅔ Protein • ⅔ Bread • 14 Optional Calories

110 Calories • 2 gm Fat • 6 gm Protein • 17 gm Carbohydrate • 296 mg Sodium • 103 mg Calcium • 2 gm Fiber

DIABETIC: 1 Starch • ½ Meat

Southern Coleslaw

I think I've probably invented enough coleslaw recipes to serve a different one every single day for the whole summer! This one mingles the tangy surprise of Tabasco with the perk-up of celery seed. After you try it, don't blame me if you wave to your friends and drawl, "Y'all better taste this!" ☺ Serves 6 (⅔ cup)

¼ cup skim milk
Sugar substitute to equal 2 tablespoons sugar
1 tablespoon dried onion flakes
4 cups purchased coleslaw mix
1 recipe JO's Sour Cream
⅓ cup Kraft fat-free mayonnaise
1 to 2 drops Tabasco sauce
¼ teaspoon celery seed

In a large bowl, combine skim milk, sugar substitute, and onion flakes. Add coleslaw mix. Mix well to combine. Cover and refrigerate for at least 1 hour. In a medium bowl, combine **JO's Sour Cream** and mayonnaise. Stir in Tabasco sauce and celery seed. Add dressing mixture to coleslaw mixture. Mix well to combine. Cover and refrigerate for at least 30 minutes. Gently stir again just before serving.

HINT: 3½ cups shredded cabbage and ½ cup shredded carrots
       may be used in place of coleslaw mix.

Each serving equals:

HE: 1⅓ Vegetable • ⅓ Skim Milk •
15 Optional Calories

56 Calories • 0 gm Fat • 4 gm Protein •
10 gm Carbohydrate • 170 mg Sodium •
140 mg Calcium • 1 gm Fiber

DIABETIC: 2 Vegetable

Ranch Corn and Tomato Salad

Before I stirred up this pretty salad, I closed my eyes and imagined the glorious colors of a California harvest! This quick and easy recipe shines with flavor, color, and crunch, and it's perfect for taking to a potluck. ☻ Serves 6 (½ cup)

2 cups frozen corn, thawed
1¼ cups diced fresh tomato
½ cup diced green bell pepper
¼ cup diced onion
¾ cup (3 ounces) shredded Kraft reduced-fat Cheddar cheese
2 tablespoons finely chopped fresh parsley or 2 teaspoons
    dried parsley flakes
⅓ cup Kraft Fat Free Ranch Dressing
2 tablespoons Kraft fat-free mayonnaise
Sugar substitute to equal 1 teaspoon sugar
¼ teaspoon black pepper

In a medium bowl, combine corn, tomato, green pepper, onion, Cheddar cheese, and parsley. In a small bowl, combine Ranch dressing, mayonnaise, sugar substitute, and black pepper. Add dressing mixture to corn mixture. Mix well to combine. Cover and refrigerate for at least 30 minutes. Gently stir again just before serving.

HINT: Thaw corn by placing in a colander and rinsing under hot water for one minute.

Each serving equals:

HE: ⅔ Bread • ⅔ Vegetable • ⅔ Protein •
¼ Slider • 5 Optional Calories

118 Calories • 2 gm Fat • 6 gm Protein •
19 gm Carbohydrate • 275 mg Sodium •
98 mg Calcium • 2 gm Fiber

DIABETIC: 1 Starch • ½ Meat

Cauliflower Party Salad

It's amazing how just a little cheese—even the reduced-fat kind—tastes like a lot when it's stirred into a crunchy pea salad. This is an excellent choice to accompany one of my cozy casseroles.

⏺ Serves 6 (¾ cup)

> 2 cups chopped fresh cauliflower
> 1½ cups frozen peas, thawed
> ¾ cup chopped celery
> ¼ cup chopped onion
> ¾ cup (3 ounces) shredded Kraft reduced-fat Cheddar cheese
> ½ cup Kraft fat-free mayonnaise
> ¼ cup Kraft Fat Free Ranch Dressing
> 1 teaspoon dried parsley flakes
> Sugar substitute to equal 2 teaspoons sugar

In a large bowl, combine cauliflower, peas, celery, onion, and Cheddar cheese. In a small bowl, combine mayonnaise, Ranch dressing, parsley flakes, and sugar substitute. Add dressing mixture to cauliflower mixture. Mix gently to combine. Cover and refrigerate for at least 30 minutes. Gently stir again just before serving.

HINT: Thaw peas by placing in a colander and rinsing under hot water for one minute.

Each serving equals:

HE: 1 Vegetable • ⅔ Protein • ½ Bread •
¼ Slider • 11 Optional Calories

---

102 Calories • 2 gm Fat • 6 gm Protein •
15 gm Carbohydrate • 424 mg Sodium •
117 mg Calcium • 3 gm Fiber

DIABETIC: 1 Vegetable • ½ Starch • ½ Meat

Confetti Broccoli Salad

Well, this will surely never make Cliff's Top Ten List because of the broccoli, but I'm sure it'll win you plenty of fans. In fact, I think that its lively blend of veggies and creamy, crunchy dressing deserves a medal—gold, of course!   ☻   Serves 6 (¾ cup)

3¼ cups chopped fresh broccoli

1 cup shredded carrots

¼ cup chopped onion

¾ cup (3 ounces) shredded Kraft reduced-fat Cheddar cheese

2 tablespoons Hormel Bacon Bits

1 recipe JO's Sour Cream

¼ cup Kraft fat-free mayonnaise

Sugar substitute to equal 2 tablespoons sugar

1½ teaspoons prepared mustard

In a large bowl, combine broccoli, carrots, onion, Cheddar cheese, and bacon bits. In a small bowl, combine **JO's Sour Cream**, mayonnaise, sugar substitute, and mustard. Add dressing mixture to broccoli mixture. Mix gently to combine. Cover and refrigerate for at least 30 minutes. Gently stir again just before serving.

Each serving equals:

HE: 1½ Vegetable • ⅔ Protein • ⅓ Skim Milk • 17 Optional Calories

---

111 Calories • 3 gm Fat • 9 gm Protein • 12 gm Carbohydrate • 367 mg Sodium • 226 mg Calcium • 2 gm Fiber

---

DIABETIC: 1 Vegetable • ½ Meat • ½ Starch

Roman Green Bean Salad

Those flat frozen Italian green beans are my friend Barbara's favorite vegetable snack. This fast and fun salad will go just beautifully with any pasta meal. ☺ Serves 4 (1 cup)

¼ cup Kraft Fat Free Italian Dressing
2 tablespoons Kraft fat-free mayonnaise
2 cups (one 16-ounce can) Italian green beans,
    rinsed and drained
1¼ cups chopped fresh tomato
¼ cup finely chopped onion
¼ cup (1 ounce) sliced ripe olives
⅓ cup (1½ ounces) shredded Kraft reduced-fat
    mozzarella cheese

In a large bowl, combine Italian dressing and mayonnaise. Stir in green beans, tomato, onion, and olives. Add mozzarella cheese. Mix well to combine. Cover and refrigerate for at least 30 minutes. Gently stir again just before serving.

Each serving equals:

HE: 1¾ Vegetable • ½ Protein • ¼ Fat •
13 Optional Calories

87 Calories • 3 gm Fat • 5 gm Protein •
10 gm Carbohydrate • 362 mg Sodium •
110 mg Calcium • 4 gm Fiber

DIABETIC: 2 Vegetable • ½ Fat

Grande Green Bean Salad

Try this for supper and see if your family doesn't "jump" for joy! Yes, it does seem unusual to stir French dressing into a Mexican dish, but you'll soon see that this is one blend of foreign flavors that really works!   ☻   Serves 6 (⅔ cup)

> 4 cups (two 16-ounce cans) cut green beans,
>   rinsed and drained
> ¼ cup finely chopped onion
> ¼ cup finely chopped green bell pepper
> 1½ cups frozen whole kernel corn, thawed
> ½ cup Kraft Fat Free French Dressing
> 1 teaspoon chili seasoning
> ¾ cup (3 ounces) shredded Kraft reduced-fat Cheddar cheese

In a large bowl, combine green beans, onion, green pepper, and corn. Add French dressing. Mix well to combine. Stir in chili seasoning and Cheddar cheese. Cover and refrigerate for at least 30 minutes. Gently stir again just before serving.

HINT: Thaw corn by placing in a colander and rinsing under hot water for one minute.

Each serving equals:

HE: 1½ Vegetable • ⅔ Protein • ½ Bread • ¼ Slider • 7 Optional Calories

130 Calories • 2 gm Fat • 6 gm Protein • 22 gm Carbohydrate • 325 mg Sodium • 120 mg Calcium • 4 gm Fiber

DIABETIC: 1 Vegetable • 1 Starch • ½ Meat

Tossed Spinach Orange Salad

This is such a pretty salad, you'll probably want to serve it to good friends at a dinner party. Its unusual combination of flavors makes it a real conversation starter! ♥ Serves 4 (1 full cup)

> 4 cups torn fresh spinach leaves, stems removed
>   and discarded
> 1 cup (one 11-ounce can) mandarin oranges,
>   rinsed and drained
> 2 cups chopped fresh cauliflower
> ½ cup Kraft Fat Free Ranch Dressing
> ½ cup unsweetened orange juice
> ¼ cup Kraft fat-free mayonnaise

In a large bowl, combine spinach leaves, oranges, and cauliflower. In a small bowl, combine Ranch dressing, orange juice, and mayonnaise. Add dressing mixture to spinach mixture. Toss gently to combine. Serve at once.

Each serving equals:

HE: 3 Vegetable • ¾ Fruit • ¾ Slider

112 Calories • 0 gm Fat • 3 gm Protein •
25 gm Carbohydrate • 391 mg Sodium •
76 mg Calcium • 3 gm Fiber

DIABETIC: 2 Vegetable • 1 Fruit

Greek Feta Salad

This delicious culinary journey offers a bundle of Mediterranean sunshine in just a few bites—and without the jet lag! Its lively appearance will decorate any party meal, too.

○ Serves 4 (2 cups)

> 3 cups shredded lettuce
> 1¼ cups unpeeled chopped cucumber
> 1 cup chopped fresh tomato
> ½ cup chopped green bell pepper
> ¼ cup chopped onion
> ¾ cup (3 ounces) crumbled feta cheese
> ¼ cup (1 ounce) sliced ripe olives
> 2 tablespoons chopped fresh parsley or 2 teaspoons
>   dried parsley flakes
> ¼ cup Kraft Fat Free Ranch Dressing
> 2 tablespoons Kraft Fat Free Italian Dressing

In a large bowl, combine lettuce, cucumber, tomato, green pepper, onion, feta cheese, olives, and parsley. Mix well to combine. Cover and refrigerate until serving time. Just before serving, in a small bowl, combine Ranch dressing and Italian dressing. Pour dressing mixture over lettuce mixture. Toss gently to combine. Serve at once.

HINT: If you can't find feta cheese, use either reduced-fat Swiss cheese or any other cheese of your choice.

Each serving equals:

> HE: 3 Vegetable • 1 Protein • ¼ Fat • ¼ Slider •
> 9 Optional Calories
> _____
> 109 Calories • 5 gm Fat • 4 gm Protein •
> 12 gm Carbohydrate • 516 mg Sodium •
> 127 mg Calcium • 2 gm Fiber
> _____
> DIABETIC: 2 Vegetable • 1 Meat

Mediterranean Tomato Mozzarella Salad

Nothing could be easier than this speedy blend of rosy-ripe tomatoes and the cheese that murmurs "Italy!" with every bite! It's a terrific "go-along," but you'll also find that a double serving makes a satisfying summer lunch entree.   ◐   Serves 4 (¾ cup)

> 3 cups cherry tomatoes, quartered
> ⅓ cup (1½ ounces) shredded Kraft reduced-fat
>   mozzarella cheese
> ⅓ cup Kraft Fat Free Ranch Dressing
> 2 tablespoons Kraft Fat Free Italian Dressing
> 1 tablespoon chopped fresh parsley or 1 teaspoon
>   dried parsley flakes

In a medium bowl, combine tomatoes and mozzarella cheese. In a small bowl, combine Ranch dressing, Italian dressing, and parsley. Add dressing mixture to tomato mixture. Mix gently to combine. Cover and refrigerate for at least 15 minutes. Gently stir again just before serving.

Each serving equals:

HE: 1½ Vegetable • ½ Protein • ¼ Slider •
17 Optional Calories

90 Calories • 2 gm Fat • 4 gm Protein •
14 gm Carbohydrate • 367 mg Sodium •
76 mg Calcium • 1 gm Fiber

DIABETIC: 2 Vegetable • ½ Fat

Layered Taco Salad

This would be a terrific choice for a simple family meal—it's easily assembled, serves 6, and is as colorful as a fiesta parade! How could anyone feel deprived when they're dining like the grand marshal? (Cliff didn't even miss the meat in this tummy-pleasing salad!)

☻ Serves 6

2½ cups finely shredded lettuce

½ cup diced celery

¾ cup diced green bell pepper

¼ cup chopped green onion

1 cup shredded carrots

10 ounces (one 16-ounce can) pinto beans,
    rinsed and drained

1 cup Kraft fat-free mayonnaise

2 tablespoons skim milk

1 tablespoon taco seasoning

¾ cup (3 ounces) shredded Kraft reduced-fat
    Cheddar cheese

1 cup diced fresh tomato

⅓ cup (1½ ounces) sliced ripe olives

6 tablespoons Land O Lakes no-fat sour cream

Evenly arrange lettuce in a 9-by-13-inch pan. In a medium bowl, combine celery, green pepper, green onion, and carrots. Arrange vegetable mixture evenly over the lettuce. Layer pinto beans over vegetables. In a small bowl, combine mayonnaise, skim milk, and taco seasoning. Spread dressing mixture evenly over vegetables. Sprinkle Cheddar cheese over top. Cover and refrigerate for at least 2 hours. Just before serving, evenly arrange tomato and olives over top. Divide mixture into 6 servings. Garnish each with 1 tablespoon sour cream.

Each serving equals:

HE: 2 Vegetable • 1½ Protein • ¼ Fat • ½ Slider • 4 Optional Calories

---

171 Calories • 3 gm Fat • 9 gm Protein • 26 gm Carbohydrate • 477 mg Sodium • 161 mg Calcium • 5 gm Fiber

---

DIABETIC: 1½ Starch • 1 Vegetable • ½ Meat

Cottage Cheese Pea Salad

Here's a recipe that's calcium-rich and full of luscious tastes, so you'll have happy teeth as well as a pleasured palate! The pickle relish will almost persuade you that you're playing hooky and on a picnic. ☻ Serves 4 (1 full cup)

> 2 cups fat-free cottage cheese
> 1/3 cup Carnation Nonfat Dry Milk Powder
> 2 cups (one 16-ounce can) small peas, rinsed and drained
> 3/4 cup (3 ounces) shredded Kraft reduced-fat Cheddar cheese
> 1/4 cup sweet pickle relish
> 1/4 cup Kraft Fat Free Thousand Island Dressing

In a medium bowl, combine cottage cheese and dry milk powder. Stir in peas and Cheddar cheese. Add pickle relish and Thousand Island dressing. Mix gently to combine. Cover and refrigerate for at least 30 minutes. Gently stir again just before serving.

HINT: Attractive served on lettuce leaves.

Each serving equals:

HE: 2 Protein • 1 Bread • 1/4 Skim Milk •
1/4 Slider • 15 Optional Calories

247 Calories • 3 gm Fat • 27 gm Protein •
28 gm Carbohydrate • 733 mg Sodium •
239 mg Calcium • 5 gm Fiber

DIABETIC: 2 1/2 Meat • 1 Starch

Chicken Club Salad

I haven't done an actual survey, but I'd be willing to wager that the most popular restaurant sandwich in America is the three-decker turkey or chicken club! Here's my take on this beloved classic, in salad form and with only a touch of bread, instead of three calorie-laden slices. ❍ Serves 4 (1½ cups)

1 cup (5 ounces) diced cooked chicken breast
⅓ cup Kraft Fat Free French Dressing
¼ cup finely chopped onion
½ cup finely chopped celery
2 tablespoons Hormel Bacon Bits
3¼ cups finely shredded lettuce
¾ cup (3 ounces) shredded Kraft reduced-fat Cheddar cheese
1 cup chopped fresh tomato
2 slices reduced-calorie bread, toasted and diced

In a large bowl, combine chicken, French dressing, onion, celery, and bacon bits. Mix well to combine. Cover and refrigerate until serving time. Just before serving, add lettuce, Cheddar cheese, tomato and toasted bread. Mix gently to combine. Serve at once.

Each serving equals:

HE: 2½ Vegetable • 2¼ Protein • ¼ Bread •
½ Slider • 2 Optional Calories

206 Calories • 6 gm Fat • 20 gm Protein •
18 gm Carbohydrate • 506 mg Sodium •
172 mg Calcium • 3 gm Fiber

DIABETIC: 2 Meat • 1 Starch • 1 Free Vegetable

Mexican Rotini Salad

Rotini pasta just seems to make eating fun, don't you think? (I bet your kids will agree with me!) This tangy preparation delivers lots of good nutrition along with taste that never ends!

◑ Serves 6 (full ¾ cup)

> 2 cups cold cooked rotini pasta, rinsed and drained
> 10 ounces (one 16-ounce can) red kidney beans,
>   rinsed and drained
> 2 tablespoons dried onion flakes
> ¾ cup (3 ounces) shredded Kraft reduced-fat Cheddar cheese
> ½ cup Kraft Fat Free Catalina or French Dressing
> ¼ cup Kraft fat-free mayonnaise
> 1 teaspoon chili seasoning

In a large bowl, combine rotini pasta, kidney beans, onion flakes, and Cheddar cheese. In a small bowl, combine Catalina dressing, mayonnaise, and chili seasoning. Add dressing mixture to pasta mixture. Mix gently to combine. Cover and refrigerate for at least 30 minutes. Gently stir again just before serving.

HINT: 1½ cups uncooked rotini pasta usually cooks to about 2 cups.

Each serving equals:

> HE: 1½ Protein • ⅔ Bread • ¼ Slider •
> 13 Optional Calories
> _____
> 175 Calories • 3 gm Fat • 8 gm Protein •
> 29 gm Carbohydrate • 434 mg Sodium •
> 111 mg Calcium • 4 gm Fiber
> _____
> DIABETIC: 2 Starch • ½ Meat

Sunny-Side-Up Pasta Salad

Here's my idea of sunshine in a salad, a scrumptious combo made tangy and rich with a classic mustard dressing. Sunny color, sunny flavor, and sunny satisfaction in your tummy!

○ Serves 4 (1 cup)

> 2 cups cold cooked rotini pasta, rinsed and drained
> 1 cup (one 8-ounce can) cut green beans, rinsed and drained
> 2 tablespoons Hormel Bacon Bits
> ⅓ cup Kraft fat-free mayonnaise
> ⅓ cup Carnation Nonfat Dry Milk Powder
> 2 tablespoons water
> 1 tablespoon Dijon Country Style Mustard
> Sugar substitute to equal 1 teaspoon sugar
> 1 teaspoon dried onion flakes
> 1 tablespoon chopped fresh parsley or 1 teaspoon
>   dried parsley flakes
> 1 cup chopped fresh tomato
> 1 hard-boiled egg, chopped

In a large bowl, combine rotini pasta, green beans, and bacon bits. In a small bowl, combine mayonnaise, dry milk powder, water, mustard, sugar substitute, onion flakes, and parsley. Add mayonnaise mixture to pasta mixture. Mix well to combine. Cover and refrigerate for at least 30 minutes. Just before serving, stir in tomato and egg.

HINT: 1½ cups uncooked rotini pasta usually cooks to about 2 cups.

Each serving equals:

HE: 1 Bread • 1 Vegetable • ¼ Skim Milk •
¼ Protein (limited) • ¼ Slider • 6 Optional Calories

---

191 Calories • 3 gm Fat • 10 gm Protein •
31 gm Carbohydrate • 443 mg Sodium •
96 mg Calcium • 2 gm Fiber

DIABETIC: 1½ Starch • 1 Vegetable

Salmon Pasta Salad

Did you know that canned salmon provides one of the best non-dairy sources of calcium? Not only that, it tastes so rich and delicious. Blended here with cold pasta and some zingy crunch, it's bound to become a new family favorite.

♥ Serves 6 (¾ cup)

> 2 cups cold cooked rotini pasta, rinsed and drained
> ¼ cup chopped onion
> ¾ cup finely chopped celery
> 1 (14.5-ounce) can red salmon, drained and flaked
> ¼ cup dill pickle relish
> ½ cup Kraft fat-free mayonnaise
> ¼ teaspoon black pepper

In a large bowl, combine pasta, onion, and celery. In a medium bowl, combine salmon, pickle relish, mayonnaise, and black pepper. Add salmon mixture to pasta mixture. Mix gently to combine. Cover and refrigerate for at least 30 minutes. Gently stir again just before serving.

HINT: 1½ cups uncooked rotini pasta usually cooks to about 2 cups.

Each serving equals:

HE: 2⅓ Protein • ⅔ Bread • ⅓ Vegetable •
¼ Slider • 3 Optional Calories

184 Calories • 4 gm Fat • 16 gm Protein •
21 gm Carbohydrate • 639 mg Sodium •
159 mg Calcium • 1 gm Fiber

DIABETIC: 2½ Meat • 1 Starch • ½ Vegetable

Sweet Salads

Grandma's Apple Salad

This is a sweet and creamy version of a traditional Waldorf salad that shines the spotlight on apples, raisins, and nuts. It's a wonderful accompaniment to any entree, but especially a pork or ham meal.　❍　Serves 8 (½ cup)

> *4 cups (8 small) cored, unpeeled, and diced*
> > *Red Delicious apples*
>
> *½ cup raisins*
> *¼ cup (1 ounce) chopped walnuts*
> **1 recipe JO's Deliteful Whipped Topping**
> *1 teaspoon lemon juice*
> *½ teaspoon apple pie spice*

In a large bowl, combine apples, raisins, and walnuts. Add **JO's Deliteful Whipped Topping**, lemon juice, and apple pie spice. Mix gently to combine. Cover and refrigerate for at least 15 minutes. Gently stir again just before serving.

HINT:　To plump up raisins without "cooking," place in a glass measuring cup and microwave on HIGH for 20 seconds.

Each serving equals:

HE: 1½ Fruit • ¼ Fat • ¼ Skim Milk • ¼ Slider • 5 Optional Calories

122 Calories • 2 gm Fat • 3 gm Protein • 23 gm Carbohydrate • 37 mg Sodium • 88 mg Calcium • 1 gm Fiber

DIABETIC: 1½ Fruit • ½ Fat

"Candy Bar" Apple Salad

This sweet salad is outrageously delicious, and filled with what I call "real food" ingredients that most dieters never expected to enjoy again! Can you really lose weight and improve your health while munching on marshmallows, peanuts, chocolate chips, and even bits of toffee? I'm living proof that it's true!

❍ Serves 8 (full ½ cup)

> 3 cups (6 small) cored, unpeeled, and diced
>   Red Delicious apples
> ½ cup (1 ounce) miniature marshmallows
> ¼ cup (1 ounce) chopped dry-roasted peanuts
> 2 tablespoons (½ ounce) Heath Toffee Bits
> 2 tablespoons (½ ounce) mini chocolate chips
> **1 recipe JO's Deliteful Whipped Topping**

In a large bowl, combine apples, marshmallows, peanuts, toffee bits, and chocolate chips. Add **JO's Deliteful Whipped Topping**. Mix gently to combine. Cover and refrigerate for at least 30 minutes. Gently stir again just before serving.

Each serving equals:

HE: ¾ Fruit • ¼ Skim Milk • ¼ Fat • ½ Slider • 10 Optional Calories

---

107 Calories • 3 gm Fat • 3 gm Protein • 19 gm Carbohydrate • 40 mg Sodium • 84 mg Calcium • 1 gm Fiber

---

DIABETIC: 1 Fruit • ½ Fat • ½ Starch/Carbohydrate

Z-Heart's Banana Salad

I invented this scrumptious salad to delight my beautiful grandson Zach, who lets me call him Z-Heart. Do your children adore bananas and pineapple as much as he does? Then win their hearts again and again with this super-sweet treat!

● Serves 6 (⅔ cup)

> 1 (4-serving) package JELL-O sugar-free instant banana
>   pudding mix
> **2 recipes JO's Evaporated Skim Milk**
> 1 cup (one 8-ounce can) crushed pineapple, packed in
>   fruit juice, undrained
> ½ cup Cool Whip Free
> 2 cups (2 medium) diced bananas
> ½ cup (1 ounce) miniature marshmallows

In a large bowl, combine dry pudding mix, **JO's Evaporated Skim Milk**, and undrained pineapple. Mix well using a wire whisk. Blend in Cool Whip Free. Add bananas and marshmallows. Mix gently to combine. Cover and refrigerate for at least 30 minutes. Gently stir again just before serving.

HINT: To prevent bananas from turning brown, mix with 1 teaspoon lemon juice or sprinkle with Fruit Fresh.

Each serving equals:

> HE: 1 Fruit • ⅓ Skim Milk • ¼ Slider •
> 15 Optional Calories
>
> ---
>
> 140 Calories • 0 gm Fat • 3 gm Protein •
> 32 gm Carbohydrate • 274 mg Sodium •
> 101 mg Calcium • 1 gm Fiber
>
> ---
>
> DIABETIC: 1 Fruit • 1 Starch/Carbohydrate

Pink Cloud Salad

If you're looking for a luscious sweet salad for a ladies' luncheon or card club meeting, look no further! Invite your friends to join you "up in the clouds," which is definitely where a taste of this delightful dish will send you.   ☯   Serves 8 (¾ cup)

1 (4-serving) package JELL-O sugar-free raspberry gelatin

1 (4-serving) package JELL-O sugar-free orange gelatin

2 cups boiling water

3 cups frozen unsweetened red raspberries

1 recipe JO's Deliteful Whipped Topping

In a large bowl, combine dry raspberry and orange gelatins. Add boiling water. Mix well to dissolve gelatin. Stir in frozen raspberries. Refrigerate for about 30 minutes or until gelatin is partially set. Blend in **JO's Deliteful Whipped Topping**. Cover and refrigerate for at least 30 minutes. Gently stir again just before serving.

Each serving equals:

HE: ½ Fruit • ¼ Skim Milk • ¼ Slider •
8 Optional Calories

68 Calories • 0 gm Fat • 4 gm Protein •
13 gm Carbohydrate • 90 mg Sodium •
87 mg Calcium • 2 gm Fiber

DIABETIC: 1 Fruit

Angel Dew Salad

There's something fun about a salad or pie that's layered, almost as if you're getting three scrumptious tastes for the "price" of one! This inventive blend of fruit, gelatin, and rich cream cheese would make a heavenly choir stand up and sing!   ☻   Serves 8

1 (4-serving) package JELL-O sugar-free orange gelatin
1 (4-serving) package JELL-O sugar-free lemon gelatin
1¼ cups boiling water
2⅓ cups Diet Mountain Dew☆
1 cup (one 8-ounce can) crushed pineapple, packed in
    fruit juice, undrained
1 cup (one 11-ounce can) mandarin oranges,
    rinsed and drained
2 cups (2 medium) diced bananas
1 (4-serving) package JELL-O sugar-free instant vanilla
    pudding mix
⅔ cup Carnation Nonfat Dry Milk Powder
1 (8-ounce) package Philadelphia fat-free cream cheese
1 cup Cool Whip Free
1 teaspoon coconut extract
½ cup (1 ounce) miniature marshmallows
2 tablespoons flaked coconut

In a large bowl, combine dry orange and lemon gelatins. Add boiling water. Mix well to dissolve gelatin. Stir in 1 cup Diet Mountain Dew. Add undrained pineapple, mandarin oranges, and bananas. Mix well to combine. Pour mixture into an 8-by-8-inch dish. Refrigerate for at least 3 hours or until firm. In a medium bowl, combine dry pudding mix and dry milk powder. Add remaining 1⅓ cups Diet Mountain Dew. Mix well using a wire whisk. Evenly spread mixture over firm gelatin mixture. Refrigerate while preparing topping. In a medium bowl, stir cream cheese with a spoon until soft. Add Cool Whip Free and coconut extract. Mix gently to combine. Fold in marshmallows. Spread topping mixture evenly over pudding layer. Sprinkle coconut evenly over top. Refrigerate for at least 30 minutes. Cut into 8 servings.

HINT: To prevent bananas from turning brown, mix with 1 teaspoon lemon juice or sprinkle with Fruit Fresh.

Each serving equals:

HE: 1 Fruit • ½ Protein • ¼ Skim Milk • ½ Slider • 7 Optional Calories

---

156 Calories • 0 gm Fat • 8 gm Protein • 31 gm Carbohydrate • 439 mg Sodium • 80 mg Calcium • 1 gm Fiber

---

DIABETIC: 1 Fruit • 1 Starch/Carbohydrate • ½ Meat

Martha Washington Salad

This creation was inspired by a favorite family recipe that's been enjoyed for years, but my version celebrates the memory—without the excess fat and sugar! Any First Lady would be proud to call it hers (and serve it on that fancy china)—especially the first one!

❂ Serves 8 (full ½ cup)

> 1 cup (one 8-ounce can) crushed pineapple, packed in
>   fruit juice, drained and ¼ cup liquid reserved
> 2 cups (one 16-ounce can) tart red cherries, packed in water,
>   drained and ¼ cup liquid reserved
> 1⅔ cups Carnation Nonfat Dry Milk Powder☆
> ½ cup Sugar Twin or Sprinkle Sweet
> ¾ cup Yoplait plain fat-free yogurt
> 1 teaspoon coconut extract
> 4 to 5 drops red food coloring
> ¾ cup Cool Whip Free
> ½ cup (1 ounce) miniature marshmallows

In a 2-cup glass measuring cup, combine reserved liquids and 1⅓ cups dry milk powder. Mix well to combine. Cover and microwave on HIGH (100% power) for 45 seconds or until mixture almost starts to boil. Stir in Sugar Twin. Re-cover and refrigerate for at least 30 minutes. In a large bowl, combine yogurt and remaining ⅓ cup dry milk powder. Stir in coconut extract, red food coloring, and Cool Whip Free. Add cooled milk mixture. Mix gently to combine. Stir in pineapple, marshmallows, and cherries. Cover and refrigerate for at least 1 hour. Gently stir again just before serving.

Each serving equals:

HE: ¾ Fruit • ¾ Skim Milk • ¼ Slider •
3 Optional Calories

---

124 Calories • 0 gm Fat • 7 gm Protein •
24 gm Carbohydrate • 103 mg Sodium •
227 mg Calcium • 1 gm Fiber

DIABETIC: 1 Fruit • 1 Skim Milk

Pineapple Grape Salad

I've always loved the look of red grapes when I find them in the market, but many recipes call for green ones instead. I wanted to create a delectable salad that gave this lovely fruit star billing. The marshmallows and pecans will persuade your taste buds that it's extra-special. ◑ Serves 6 (¾ cup)

> 1 (4-serving) package JELL-O sugar-free vanilla
>   cook-and-serve pudding mix
> ⅔ cup Carnation Nonfat Dry Milk Powder
> 1 cup (one 8-ounce can) crushed pineapple, packed in
>   fruit juice, undrained
> 1 cup water
> ¾ cup Cool Whip Free
> ½ cup (1 ounce) miniature marshmallows
> 2 tablespoons (½ ounce) chopped pecans
> 2 cups (12 ounces) seedless red grapes

In a medium saucepan, combine dry pudding mix, dry milk powder, undrained pineapple, and water. Cook over medium heat for 10 minutes or until mixture thickens and starts to boil, stirring constantly. Remove from heat. Place saucepan on a wire rack and allow to cool for 30 minutes. Pour mixture into a large bowl. Blend in Cool Whip Free. Add marshmallows, pecans, and grapes. Mix gently to combine. Cover and refrigerate for at least 30 minutes. Gently stir again just before serving.

Each serving equals:

HE: 1 Fruit • ⅓ Fat • ⅓ Skim Milk • ½ Slider • 2 Optional Calories

154 Calories • 2 gm Fat • 3 gm Protein • 31 gm Carbohydrate • 126 mg Sodium • 105 mg Calcium • 1 gm Fiber

DIABETIC: 1 Fruit • 1 Starch/Carbohydrate

Sherry's Cranberry Salad

When those bags of cranberries are piled high, especially during the holiday months, I can almost taste their tangy sweetness! I created this gorgeous dish by special request (the original version packed 33 grams of fat per serving), and everyone who's tried it agrees it's as tasty as it is pretty . . . almost a cheesecake salad.

● Serves 8 (¾ cup)

> 2 cups finely chopped fresh or frozen cranberries
> ½ cup Sugar Twin or Sprinkle Sweet
> 1 (8-ounce) package Philadelphia fat-free cream cheese
> **1 recipe JO's Deliteful Whipped Topping**
> ⅔ cup (2 medium) mashed bananas
> 1 teaspoon lemon juice
> ½ cup (2 ounces) chopped walnuts
> 1 cup (2 ounces) miniature marshmallows

In a medium bowl, combine cranberries and Sugar Twin. Cover and refrigerate for 4 hours or up to overnight. In a medium bowl, stir cream cheese with a spoon until soft. Add **JO's Deliteful Whipped Topping**. Mix well to combine. Stir in bananas and lemon juice. Add walnuts and marshmallows. Mix gently to combine. Stir in cranberry mixture. Cover and refrigerate for at least 30 minutes. Gently stir again just before serving.

Each serving equals:

> HE: ¾ Fruit • ¾ Protein • ¼ Fat • ¼ Slider • 10 Optional Calories
>
> ---
>
> 152 Calories • 4 gm Fat • 8 gm Protein • 21 gm Carbohydrate • 226 mg Sodium • 121 mg Calcium • 1 gm Fiber
>
> ---
>
> DIABETIC: 1 Fruit • 1 Starch/Carbohydrate • ½ Meat • ½ Fat

Caribbean Party Salad

Most people think of graham crackers as plain and just a little sweet, but when you stir them into this tropical delight, you'll be deliciously pleased at the culinary pleasure they add to this vivid dish. ☻ Serves 8 (¾ cup)

1 (4-serving) package JELL-O sugar-free instant banana
    pudding mix
1 cup Carnation Nonfat Dry Milk Powder☆
1 cup (one 8-ounce can) crushed pineapple, packed in
    fruit juice, drained and ⅓ cup liquid reserved
1 cup water
¾ cup Yoplait plain fat-free yogurt
1 teaspoon coconut extract
Sugar substitute to equal 2 tablespoons sugar
¾ cup Cool Whip Free
1 cup (one 11-ounce can) mandarin oranges,
    rinsed and drained
12 (2½-inch) graham crackers, broken into large pieces

In a large bowl, combine dry pudding mix and ⅔ cup dry milk powder. Add reserved pineapple juice and water. Mix well using a wire whisk. In a small bowl, combine yogurt and remaining ⅓ cup dry milk powder. Stir in coconut extract and sugar substitute. Blend in Cool Whip Free. Add yogurt mixture to pudding mixture. Mix gently until well blended. Fold in pineapple and mandarin oranges. Cover and refrigerate for at least 30 minutes. Just before serving, stir in graham cracker pieces.

Each serving equals:

HE: ½ Skim Milk • ½ Fruit • ½ Bread •
¼ Slider • 5 Optional Calories

141 Calories • 1 gm Fat • 5 gm Protein •
28 gm Carbohydrate • 321 gm Sodium •
155 mg Calcium • 0 gm Fiber

DIABETIC: 1 Starch • ½ Skim Milk • ½ Fruit

Frog Eye Salad

This is a real old-timey recipe and a flavorful Midwest tradition. Don't let the funny name discourage you from trying it! It combines a wonderful variety of tastes and textures into a creamy, pretty salad. Nelda, mother of my daughter-in-law, Pam, made her version of this for Pam and James's wedding reception, and I was inspired!

☻ Serves 8 (⅔ cup)

> 1 (4-serving) package JELL-O sugar-free vanilla
>   cook-and-serve pudding mix
> ⅔ cup Carnation Nonfat Dry Milk Powder
> 2 cups (two 8-ounce cans) crushed pineapple, packed in
>   fruit juice, drained, and ½ cup liquid reserved
> 1 cup water
> 1 teaspoon vanilla extract
> 2 cups cold cooked acini de pepe macaroni,
>   rinsed and drained
> 1 cup (one 11-ounce can) mandarin oranges,
>   rinsed and drained
> ¾ cup Cool Whip Free
> ½ cup (1 ounce) miniature marshmallows

In a large saucepan, combine dry pudding mix and dry milk powder. Add reserved pineapple juice and water. Mix well to combine. Cook over medium heat for 10 minutes or until mixture thickens and starts to boil, stirring constantly. Remove from heat. Stir in vanilla extract and macaroni. Place saucepan on a wire rack and allow to cool for 10 minutes. Fold in mandarin oranges and pineapple. Pour mixture into a large bowl. Cover and refrigerate for at least 1 hour. Stir in Cool Whip Free and marshmallows. Re-cover and refrigerate for at least 20 minutes. Gently stir again just before serving.

HINTS: 1. If you can't find acini de pepe macaroni, use the small-
est macaroni available.
2. A scant 1 cup uncooked acini de pepe macaroni usually cooks to about 2 cups.

Each serving equals:

HE: ¾ Fruit • ½ Bread • ¼ Skim Milk • ¼ Slider •
7 Optional Calories

148 Calories • 0 gm Fat • 4 gm Protein •
33 gm Carbohydrate • 96 mg Sodium •
84 mg Calcium • 1 gm Fiber

DIABETIC: 1 Fruit • 1 Starch/Carbohydrate

Peach Melba Cottage Cheese Salad

Inspired by singer Nellie Melba, the dessert called Peach Melba has been a restaurant classic for decades. I created this peach and raspberry splendor with my daughter, Becky, in mind (she loves this combo). It blends those delightful fruits with creamy cottage cheese and yogurt. Don't be surprised if friends and family start to "sing" your praises!   ☻   Serves 8 (¾ cup)

> 2 cups fat-free cottage cheese
> 1 (4-serving) package JELL-O sugar-free raspberry gelatin
> ¾ cup Yoplait plain fat-free yogurt
> ⅓ cup Carnation Nonfat Dry Milk Powder
> ½ cup Cool Whip Free
> 2 cups (one 16-ounce can) sliced peaches, packed in
>   fruit juice, drained, and coarsely chopped
> 1½ cups fresh red raspberries

In a large bowl, combine cottage cheese and dry gelatin. Mix well to combine. In a small bowl, combine yogurt and dry milk powder. Stir in Cool Whip Free. Add yogurt mixture to cottage cheese mixture. Mix well to combine. Fold in chopped peaches and raspberries. Cover and refrigerate for at least 30 minutes. Gently stir again just before serving.

HINT: Unsweetened frozen raspberries may be used; just be sure to thaw and drain well before using.

Each serving equals:

HE: ¾ Fruit • ½ Protein • ¼ Skim Milk •
12 Optional Calories

112 Calories • 0 gm Fat • 11 gm Protein •
17 gm Carbohydrate • 274 mg Sodium •
110 mg Calcium • 2 gm Fiber

DIABETIC: 1 Fruit • 1 Meat

Hawaiian Cottage Fruit Salad

You just can't have too many summer salads, and this is one of my recent favorites. It's cool and smooth and richly flavored with the sweetest citrus you can find. For all of you who adore pineapple and the fun flavor of mandarin oranges, this will be a popular addition to your repertoire. ☻   Serves 8 (⅔ cup)

2 cups fat-free cottage cheese
2 cups (two 8-ounce cans) crushed pineapple, packed in
    fruit juice, drained
1 (4-serving) package JELL-O sugar-free orange gelatin
1 cup (one 11-ounce can) mandarin oranges,
    rinsed and drained
¾ cup Yoplait plain fat-free yogurt
⅓ cup Carnation Nonfat Dry Milk Powder
¾ cup Cool Whip Free

In a large bowl, combine cottage cheese and pineapple. Add dry gelatin. Mix well to combine. Fold in mandarin oranges. In a small bowl, combine yogurt and dry milk powder. Stir in Cool Whip Free. Add yogurt mixture to cottage cheese mixture. Mix gently to combine. Cover and refrigerate for at least 30 minutes. Gently stir again just before serving.

Each serving equals:

HE: ¾ Fruit • ½ Protein • ¼ Skim Milk •
16 Optional Calories

128 Calories • 0 gm Fat • 11 gm Protein •
21 gm Carbohydrate • 274 mg Sodium •
113 mg Calcium • 0 gm Fiber

DIABETIC: 1 Fruit • 1 Meat • ½ Starch/Carbohydrate

Main Dishes

JO's Fettuccine Alfredo

Did you skip my special dairy mix section and head straight for the recipes? Well, take a quick leap back and stir up my **JO's Evaporated Skim Milk** before you prepare this dreamy, creamy Italian treat that's no longer off-limits to healthy gourmets.

❍ Serves 4 (¾ cup)

> *1 recipe JO's Evaporated Skim Milk*
> *1 tablespoon + 1 teaspoon reduced-calorie margarine*
> *1 teaspoon dried parsley flakes*
> *½ cup (1½ ounces) grated Kraft fat-free Parmesan cheese*
> *2 cups hot cooked fettuccine, rinsed and drained*

In a large skillet, combine **JO's Evaporated Skim Milk**, margarine, and parsley flakes. Cook over medium heat for 3 to 4 minutes or until mixture is heated through. Add Parmesan cheese and hot fettuccine. Mix well to combine. Continue cooking for 1 to 2 minutes, stirring constantly. Serve at once.

HINT: 1½ cups uncooked fettuccine usually cooks to about 2 cups.

Each serving equals:

HE: 1 Bread • ½ Fat • ½ Protein • ¼ Skim Milk

149 Calories • 1 gm Fat • 6 gm Protein •
29 gm Carbohydrate • 209 mg Sodium •
74 mg Calcium • 2 gm Fiber

DIABETIC: 1½ Starch • ½ Meat • ½ Fat

Cheese Sandwich Bake ❄

Remember your school days when your favorite lunchtime treat was grilled-cheese sandwiches? Well, I've taken that unforgettable classic and given it a luscious update that could be nicknamed "Grilled Cheese French Toast"! ☻ Serves 6

> 12 slices reduced-calorie white bread
> 8 (¾-ounce) slices Kraft reduced-fat American cheese
> 4 eggs or equivalent in egg substitute
> 2 cups skim milk
> ½ teaspoon lemon pepper

Spray a 9-by-13-inch baking dish with butter-flavored cooking spray. Arrange 6 slices of bread in prepared baking dish. Top each with 1 slice of cheese and another slice of bread. In a medium bowl, beat eggs with a fork until fluffy. Add skim milk and lemon pepper. Mix well to combine. Pour egg mixture evenly over sandwiches. Cut remaining 2 slices of cheese into 3 pieces each. Place a piece of cheese on top of each sandwich. Cover and refrigerate for at least 2 hours or up to 24 hours. Preheat oven to 350 degrees. Bake for 40 minutes or until top is puffed and golden brown. Place baking dish on a wire rack and let set for 5 minutes. Divide into 6 servings.

Each serving equals:

HE: 2 Protein (⅔ limited) • 1 Bread • ⅓ Skim Milk

240 Calories • 8 gm Fat • 17 gm Protein •
25 gm Carbohydrate • 760 mg Sodium •
314 mg Calcium • 5 gm Fiber

DIABETIC: 2 Meat • 1½ Starch

Impossible Zucchini Tomato Pie

I'm such an optimist, the word "impossible" only makes me more determined to find a way to pull off what seems too hard to do! This recipe begins by emerging from your blender in liquid form but, in what seems only a short time, turns into a magically satisfying main-dish pie. A great idea when your fridge is full of end-of-summer zucchini!   ☻   Serves 6

2½ cups unpeeled chopped zucchini
1 cup peeled and chopped fresh tomatoes
½ cup chopped onion
¼ cup (¾ ounce) grated Kraft fat-free Parmesan cheese
1 recipe JO's Buttermilk
¾ cup Bisquick Reduced Fat Baking Mix
2 eggs or equivalent in egg substitute
1 teaspoon Italian seasoning

Preheat oven to 375 degrees. Spray a deep-dish 10-inch pie plate with olive oil–flavored cooking spray. Layer zucchini, tomatoes, and onion in prepared pie plate. Sprinkle Parmesan cheese evenly over the top. In a blender container, combine **JO's Buttermilk**, baking mix, eggs, and Italian seasoning. Cover and process on HIGH for 30 seconds or until mixture is smooth. Pour mixture over vegetables and cheese. Bake for 35 to 40 minutes or until a knife inserted in the center comes out clean. Place pie plate on a wire rack and let set for 5 minutes. Cut into 6 servings.

Each serving equals:

HE: 1⅓ Vegetable • ⅔ Bread •
½ Protein (⅓ limited) • ⅓ Skim Milk

139 Calories • 3 gm Fat • 7 gm Protein •
21 gm Carbohydrate • 294 mg Sodium •
125 mg Calcium • 2 gm Fiber

DIABETIC: 1 Vegetable • 1 Meat • 1 Starch

Broccoli Noodle Casserole

This tastes every bit as good as the dish my grandmother used to take out of her wood-burning cookstove all those years ago. The first bite brings me back to those memorable days when I was a little girl helping at her boarding house, and now I can enjoy those cozy memories as often as I please. ☕ Serves 6

1 (16-ounce) package frozen chopped broccoli
½ cup finely chopped onion
2¼ cups (3¾ ounces) uncooked noodles
3 cups hot water
1½ cups (one 12-fluid-ounce can) Carnation Evaporated
    Skim Milk
3 tablespoons all-purpose flour
5 (¾ ounce) slices Kraft reduced-fat Swiss cheese,
    shredded
½ cup (one 2.5-ounce jar) sliced mushrooms, drained
¼ teaspoon lemon pepper
¼ cup (¾ ounce) grated Kraft fat-free Parmesan cheese

Preheat oven to 325 degrees. Spray an 8-by-8-inch baking dish with butter-flavored cooking spray. In a large saucepan, cook broccoli, onion, and noodles in water for 10 minutes or until just tender. Drain. Return vegetables to saucepan. Meanwhile, in a covered jar, combine evaporated skim milk and flour. Shake well to blend. Pour milk mixture into a medium saucepan sprayed with butter-flavored cooking spray. Add Swiss cheese, mushrooms, and lemon pepper. Mix well to combine. Cook over medium heat for 5 minutes or until mixture thickens and cheese melts, stirring often. Remove from heat. Add milk mixture to drained broccoli mixture. Mix well to combine. Pour mixture into prepared baking dish. Evenly sprinkle Parmesan cheese over top. Bake for 25 to 30 minutes. Place baking dish on a wire rack and let set for 5 minutes. Divide into 6 servings.

Each serving equals:

HE: 1⅓ Vegetable • 1 Bread • 1 Protein •
½ Skim Milk

---

190 Calories • 2 gm Fat • 13 gm Protein •
30 gm Carbohydrate • 233 mg Sodium •
294 mg Calcium • 4 gm Fiber

DIABETIC: 1½ Starch • 1 Vegetable • 1 Meat

Rio Grande Cabbage Bake

Take a quick look at the exchanges for this recipe, and I bet I can hear you gasp from out here in DeWitt! This rich and tangy casserole delivers more than 500 milligrams of calcium in just one hearty serving—and packs even more flavor punch! My son James didn't care about that, only that it tasted great.   ☻   Serves 4

> 3 cups purchased coleslaw mix
> 1½ cups hot water
> 1½ cups (one 12-fluid-ounce can) Carnation Evaporated
>   Skim Milk
> 3 tablespoons all-purpose flour
> ¾ cup (3 ounces) shredded Kraft reduced-fat Cheddar cheese
> ½ cup chunky salsa (mild, medium, or hot)
> 1½ cups hot cooked rice
> 1½ cups (9 ounces) diced Dubuque 97% fat-free ham or
>   any extra-lean ham

Preheat oven to 350 degrees. Spray an 8-by-8-inch baking dish with butter-flavored cooking spray. In a medium saucepan, cook coleslaw mix in water for about 10 minutes or just until tender. Drain. In a covered jar, combine evaporated skim milk and flour. Shake well to blend. Pour mixture into a large saucepan sprayed with butter-flavored cooking spray. Cook over medium heat for 5 minutes or until mixture thickens, stirring often. Stir in Cheddar cheese. Add salsa, rice, and ham. Mix well to combine. Stir in drained coleslaw mix. Pour mixture into prepared baking dish. Bake for 30 minutes. Place baking dish on a wire rack and let set for 5 minutes. Divide into 4 servings.

HINTS   1. 2¼ cups shredded cabbage and ¾ cup shredded carrots may be used in place of purchased coleslaw mix.
　　　　　2. 1 cup uncooked rice usually cooks to about 1½ cups.

Each serving equals:

HE: 2½ Protein • 1¾ Vegetable • 1 Bread •
¾ Skim Milk

290 Calories • 6 gm Fat • 25 gm Protein •
34 gm Carbohydrate • 806 mg Sodium •
501 mg Calcium • 2 gm Fiber

DIABETIC: 2 Meat • 1 Starch • 1 Vegetable •
1 Skim Milk

Corn and Rice Bake ❄

When it's cold outside and you're swamped with holiday preparations, why not stir up this cozy blend of cheese, rice, and corn, slip it into the oven, and get on with your tree trimming. In an hour, it'll be time to warm your insides just as holiday decorating always warms your heart.　○　Serves 6

1½ cups finely chopped celery
½ cup chopped onion
1 cup hot cooked rice
2 cups frozen whole kernel corn, thawed
⅔ cup Carnation Nonfat Dry Milk Powder
¾ cup water
1½ cups (6 ounces) shredded Kraft reduced-fat
　　Cheddar cheese
1 teaspoon dried parsley flakes
¼ teaspoon lemon pepper

Preheat oven to 325 degrees. Spray an 8-by-8-inch baking dish with butter-flavored cooking spray. In a large skillet sprayed with butter-flavored cooking spray, sauté celery and onion for 10 minutes or until tender. Stir in rice and corn. In a covered jar, combine dry milk powder and water. Shake well to blend. Add milk mixture to celery mixture. Mix well to combine. Stir in Cheddar cheese, parsley flakes, and lemon pepper. Lower heat and simmer for 2 to 3 minutes. Pour mixture into prepared baking dish. Cover and bake for 45 minutes. Uncover and bake for an additional 10 minutes. Place baking dish on a wire rack and let set for 5 minutes. Divide into 6 servings.

HINTS:　1. ⅔ cup uncooked rice usually cooks to about 1 cup.
　　　　　2. Thaw corn by placing in a colander and rinsing under hot water for one minute.

Each serving equals:

HE: 1⅓ Protein • 1 Bread • ⅔ Vegetable •
⅓ Skim Milk

193 Calories • 5 gm Fat • 13 gm Protein •
24 gm Carbohydrate • 331 mg Sodium •
315 mg Calcium • 2 gm Fiber

DIABETIC: 1½ Starch • 1 Meat

California Vegetable Biscuit Casserole

Here's a rich and creamy dish that is so easy to fix, and looks and smells so good, your family will race to the table when it's on the menu. Adding ham is a great way to vary the dish once it becomes a regular at your house. ● Serves 6

> 1 (16-ounce) package frozen carrot, broccoli, and
>   cauliflower blend
> 2½ cups water☆
> 1 (10¾-ounce) can Healthy Request Cream of
>   Mushroom Soup
> ⅔ cup Carnation Nonfat Dry Milk Powder
> ½ cup finely chopped onion
> ½ cup (one 2.5-ounce jar) sliced mushrooms,
>   rinsed and drained
> ¼ teaspoon black pepper
> 1 teaspoon Italian seasoning
> 1½ cups (6 ounces) shredded Kraft reduced-fat
>   Cheddar cheese☆
> 1 (7.5-ounce) can Pillsbury refrigerated buttermilk biscuits

Preheat oven to 375 degrees. Spray an 8-by-8-inch baking dish with olive oil–flavored cooking spray. In a large saucepan, cook vegetables in 2 cups water for 10 minutes or just until tender. Drain. In a large bowl, combine mushroom soup, dry milk powder, and remaining ½ cup water. Add onion, mushrooms, black pepper, Italian seasoning, and 1 cup Cheddar cheese. Mix well to combine. Stir in drained vegetables. Pour mixture into prepared baking dish. Separate and cut each biscuit into 4 pieces. Evenly sprinkle biscuit pieces over vegetable mixture. Bake for 20 minutes. Sprinkle remaining ½ cup Cheddar cheese evenly over top. Continue baking for 15 minutes. Place baking dish on a wire rack and let set for 5 minutes. Divide into 6 servings.

HINTS: 1. 3 cups frozen vegetables of your choice may be used in place of blended vegetables.
2. 1½ cups (9 ounces) finely diced Dubuque 97% fat-free or any extra-lean ham may be added, but be sure to add 1 Protein exchange to each serving.

Each serving equals:

HE: 1⅓ Vegetable • 1⅓ Protein • 1¼ Bread •
⅓ Skim Milk • ¼ Slider • 7 Optional Calories

277 Calories • 9 gm Fat • 16 gm Protein •
33 gm Carbohydrate • 903 mg Sodium •
375 mg Calcium • 3 gm Fiber

DIABETIC: 1½ Starch • 1 Vegetable • 1 Meat • ½ Fat

Yucatan Shepherd's Pie

Shepherd's pie always promises a dish that's topped with creamy mashed potatoes, then baked until it's golden brown and bubbling over with tasty goodness. Here's a meatless entree that's completely satisfying and can be just as spicy as you like. (Cliff likes it HOT!)

● Serves 4

½ cup chopped onion
½ cup chopped green bell pepper
10 ounces (one 16-ounce can) pinto beans, rinsed and
    drained
1¾ cups (one 15-ounce can) Hunt's Chunky Tomato Sauce
1 teaspoon chili seasoning
1¾ cups water
1⅓ cups (3 ounces) instant potato flakes
⅓ cup Carnation Nonfat Dry Milk Powder
½ cup chunky salsa (mild, medium, or hot)
1 teaspoon dried parsley flakes
¾ cup (3 ounces) shredded Kraft reduced-fat
    Cheddar cheese☆

Preheat oven to 350 degrees. Spray an 8-by-8-inch baking dish with olive oil–flavored cooking spray. In a large skillet sprayed with olive oil–flavored cooking spray, sauté onion and green pepper for 5 minutes or until tender. Stir in pinto beans, tomato sauce, and chili seasoning. Lower heat and simmer for 5 minutes, stirring occasionally. Pour mixture into prepared baking dish. In a medium saucepan, bring water to a boil. Remove from heat. Stir in potato flakes and dry milk powder. Add salsa, parsley flakes, and ½ cup Cheddar cheese. Mix gently to combine. Evenly spread potato mixture over bean mixture. Sprinkle remaining ¼ cup Cheddar cheese over top. Bake for 20 minutes. Place baking dish on a wire rack and let set for 5 minutes. Divide into 4 servings.

Each serving equals:

HE: 2½ Vegetable • 2¼ Protein • 1 Bread •
¼ Skim Milk

272 Calories • 4 gm Fat • 16 gm Protein •
43 gm Carbohydrate • 955 mg Sodium •
306 mg Calcium • 9 gm Fiber

DIABETIC: 2 Vegetable • 2 Starch • 1 Meat

Mexican Bean and Cornbread Pie

I love the moment when you cut into the crusty top of this fragrant dinner pie—and the spicy aroma nearly overwhelms you! Cornbread is a popular accompaniment to Mexican meals, but here I've baked it right into the dish. ☻ Serves 6

> 2 cups (one 16-ounce can) cut green beans, rinsed and drained
> 10 ounces (one 16-ounce can) red kidney beans, rinsed and drained
> 2 cups (one 16-ounce can) cut wax beans, rinsed and drained
> 1 cup (one 8-ounce can) Hunt's Tomato Sauce
> 1 cup chunky salsa (mild, medium, or hot)
> 3/4 cup (4.5 ounces) yellow cornmeal
> 1 tablespoon all-purpose flour
> 1 tablespoon Sugar Twin or Sprinkle Sweet
> 1½ teaspoons baking powder
> 1 teaspoon dried parsley flakes
> 1 teaspoon taco seasoning
> 2/3 cup Carnation Nonfat Dry Milk Powder
> 3/4 cup water
> 1 egg or equivalent in egg substitute

Preheat oven to 375 degrees. Spray a 10-inch deep-dish pie plate with olive oil–flavored cooking spray. In a large bowl, combine green beans, kidney beans, wax beans, tomato sauce, and salsa. Pour mixture into prepared pie plate. In a medium bowl, combine cornmeal, flour, Sugar Twin, baking powder, parsley flakes, and taco seasoning. In a small bowl, combine dry milk powder and water. Add egg. Mix well to combine. Stir milk mixture into cornmeal mixture. Evenly spread cornbread batter over bean mixture. Bake for 35 to 40 minutes. Place pie plate on a wire rack and let set for 5 minutes. Divide into 6 servings.

Each serving equals:

HE: 2⅓ Vegetable • 1 Bread • 1 Protein •
⅓ Skim Milk • 7 Optional Calories

181 Calories • 1 gm Fat • 9 gm Protein •
34 gm Carbohydrate • 683 mg Sodium •
259 mg Calcium • 6 gm Fiber

DIABETIC: 2 Vegetable • 1½ Starch

Macaroni Lasagna ❄

Many families are looking for easy and delicious meatless entrees that still pack lots of taste satisfaction into each hearty serving. Why not try this lively lasagna that sparkles with three—count 'em—three tasty cheeses?   ☻   Serves 4

1¾ cups (one 15-ounce can) Hunt's Chunky Tomato Sauce
⅓ cup Carnation Nonfat Dry Milk Powder
½ cup water
¼ teaspoon dried minced garlic
1 teaspoon Italian seasoning
¼ teaspoon black pepper
2 cups hot cooked macaroni, rinsed and drained
1½ cups fat-free cottage cheese
¾ cup (3 ounces) shredded Kraft reduced-fat
    mozzarella cheese
¼ cup (¾ ounce) grated Kraft fat-free
    Parmesan cheese

Preheat oven to 350 degrees. Spray an 8-by-8-inch baking dish with olive oil–flavored cooking spray. In a large skillet sprayed with olive oil–flavored cooking spray, combine tomato sauce, dry milk powder, and water. Add garlic, Italian seasoning, and black pepper. Mix well to combine. Simmer for 5 minutes. Spoon about ⅔ cup sauce mixture into prepared baking dish. Layer half of macaroni, half of cottage cheese, and half of mozzarella cheese over top. Spoon half of remaining sauce over mozzarella cheese. Repeat layer with remaining macaroni, cottage cheese, and mozzarella cheese. Spoon the remaining sauce evenly over top. Evenly sprinkle Parmesan cheese over sauce. Bake for 1 hour. Place baking dish on a wire rack and let set for 5 minutes. Cut into 4 servings.

HINT: 1⅓ cups uncooked macaroni usually cooks to about 2
      cups.

Each serving equals:

HE: 2 Protein • 1¾ Vegetable • 1 Bread •
¼ Skim Milk

276 Calories • 4 gm Fat • 25 gm Protein •
35 gm Carbohydrate • 948 mg Sodium •
266 mg Calcium • 3 gm Fiber

DIABETIC: 2½ Meat • 2 Vegetable • 1½ Starch

Leniwe Pierogi

One of the best parts of living in our nation of immigrants is the opportunity to savor the best ethnic dishes brought to the New World from the Old. These easy Polish potato dumplings (*leniwe* means "lazy") will please your family with their old-timey flavor, but I've reinvented this classic in a great low-fat version.

❍ Serves 6

½ cup chopped onion
1¾ cups (one 15-ounce can) Hunt's Chunky Tomato Sauce
2 teaspoons dried parsley flakes
¼ teaspoon black pepper
1 tablespoon Sugar Twin or Sprinkle Sweet
2¼ cups (12 ounces) peeled and diced hot cooked potatoes
1 cup fat-free cottage cheese
½ cup + 1 tablespoon all-purpose flour
½ cup + 1 tablespoon (2¼ ounces) dried fine bread crumbs
1 egg or equivalent in egg substitute
½ cup Land O Lakes no-fat sour cream

In a medium saucepan sprayed with butter-flavored cooking spray, sauté onion for 5 minutes or until tender. Add tomato sauce, parsley flakes, black pepper, and Sugar Twin. Mix well to combine. Bring mixture to a boil. Lower heat and simmer while preparing pierogi. Place potatoes in a large bowl. Mash well with a potato masher or fork. Add cottage cheese, flour, bread crumbs, and egg. Mix well to combine. Stir mixture into a ball. Cut ball in half. Place each half on waxed paper. Roll each half with hands to form a long thin "sausage" about 1 inch thick. Cut dough diagonally into pierogi about 1 inch long. (Should make about 24 pieces). Place pierogi in a large saucepan filled with boiling water. Cook pierogi for 2 to 3 minutes or just until pierogi rise to the top. (DO NOT OVERCOOK). Remove from saucepan with a slotted spoon. Repeat until all pierogi are cooked. Just before serving, stir sour cream into simmering tomato sauce mixture. For each serving, place 4 pieces, or ⅙ of pierogi, on a plate and spoon about ⅓ cup sauce mixture over top.

HINT: 1½ cups cold mashed potatoes or 1½ cups prepared instant mashed potatoes may be used in place of cooked potatoes.

Each serving equals:

HE: 1½ Bread • 1⅓ Vegetable • ½ Protein • ¼ Slider • 2 Optional Calories

201 Calories • 1 gm Fat • 11 gm Protein • 37 gm Carbohydrate • 726 mg Sodium • 76 mg Calcium • 3 gm Fiber

DIABETIC: 2 Starch • 1 Vegetable • 1 Meat

Zucchini Veggie Noodle Skillet

Don't you just love a one-dish meal that takes less than 20 minutes from start to finish? We're all so busy these days, but because we're determined to eat healthy, we need quick dishes like this skillet supper that sautés up so fast, you can probably linger over coffee.

☕ Serves 4

> 2 cups unpeeled chopped zucchini
> 1 cup shredded carrots
> ½ cup chopped onion
> 1¾ cups (one 15-ounce can) Hunt's Chunky Tomato Sauce
> ⅓ cup Carnation Nonfat Dry Milk Powder
> ½ cup (one 2.5-ounce jar) sliced mushrooms, drained
> 1 teaspoon Italian seasoning
> 1 teaspoon Sugar Twin or Sprinkle Sweet
> ¼ teaspoon black pepper
> 2 cups hot cooked noodles, rinsed and drained
> ¼ cup (¾ ounce) grated Kraft fat-free Parmesan cheese

In a large skillet sprayed with olive oil–flavored cooking spray, sauté zucchini, carrots, and onion for 10 minutes or until vegetables are just tender. Add tomato sauce, dry milk powder, mushrooms, Italian seasoning, Sugar Twin, and black pepper. Mix well to combine. Stir in noodles. Lower heat, cover, and simmer for 5 to 7 minutes, stirring occasionally. For each serving, spoon full 1 cup noodle mixture on a plate and sprinkle 1 tablespoon Parmesan cheese over top.

HINT: 1¾ cups uncooked noodles usually cooks to about 2 cups.

Each serving equals:

HE: 3¾ Vegetable • 1 Bread • ¼ Protein •
¼ Skim Milk • 1 Optional Calorie

197 Calories • 1 gm Fat • 9 gm Protein •
38 gm Carbohydrate • 811 mg Sodium •
102 mg Calcium • 6 gm Fiber

DIABETIC: 4 Vegetable • 1 Starch *or* 2½ Starch

Linguine with Tuna & Veggies ❄

If you want to include more fish in your diet, but aren't sure what your family will like, here's a scrumptious pasta concoction that's as thick and creamy as any served at your local trattoria! Peas and carrots just make it an even more popular family favorite.

● Serves 4 (1 cup)

1 (10¾-ounce) can Healthy Request Cream of Mushroom Soup
1 recipe JO's Evaporated Skim Milk
½ cup frozen peas, thawed
1 cup (one 8-ounce can) sliced carrots, rinsed and drained
1 (6-ounce) can white tuna, packed in water, drained and flaked
¼ cup (¾ ounce) grated Kraft fat-free Parmesan cheese
1½ cups hot cooked linguine, rinsed and drained
¼ teaspoon black pepper

In a large skillet sprayed with butter-flavored cooking spray, combine mushroom soup and **JO's Evaporated Skim Milk**. Stir in peas and carrots. Cook over medium heat for 5 minutes, stirring often. Add tuna, Parmesan cheese, linguine, and black pepper. Mix well to combine. Lower heat and simmer for 5 minutes or until mixture is heated through, stirring occasionally.

HINTS: 1. Thaw peas by placing in a colander and rinsing under hot water for one minute.
2. 1 cup uncooked linguine usually cooks to about 1½ cups.
3. Spaghetti may be substituted for linguine.

Each serving equals:

HE: 1 Bread • 1 Protein • ¼ Skim Milk •
¼ Vegetable • ½ Slider • 1 Optional Calorie

182 Calories • 2 gm Fat • 17 gm Protein •
24 gm Carbohydrate • 607 mg Sodium •
141 mg Calcium • 2 gm Fiber

DIABETIC: 2½ Meat • 1½ Starch

West Coast Tuna Bake

I experimented for a long time before coming up with my **JO's Dry Casserole Soup Mix**. I think you'll be surprised what a difference this homemade "convenience food" makes in your menu planning. And isn't it great that fat-free cream cheese tastes just as rich as the old-fashioned kind! ☻ Serves 4

1 (6-ounce) can white tuna, packed in water,
    drained and flaked
1½ cups hot cooked garden-variety rotini pasta,
    rinsed and drained
½ cup frozen peas, thawed
¼ cup (1 ounce) sliced pimiento-stuffed olives
2 tablespoons dried onion flakes
⅓ cup JO's Dry Casserole Soup Mix
1 cup water
¼ teaspoon lemon pepper
½ cup (4 ounces) Philadelphia fat-free cream cheese

Preheat oven to 350 degrees. Spray an 8-by-8-inch baking dish with olive oil–flavored cooking spray. In a large bowl, combine tuna, rotini pasta, peas, olives, and onion flakes. In a small bowl, combine **JO's Dry Casserole Soup Mix**, water, and lemon pepper. Add cream cheese. Mix well until blended. Stir mixture into tuna mixture. Spread mixture into prepared baking dish. Bake for 35 to 40 minutes. Place baking dish on a wire rack and let set for 5 minutes. Divide into 4 servings.

HINTS: 1. 1 cup uncooked rotini pasta usually cooks to about 1½ cups.
        2. Thaw peas by placing in a colander and rinsing under hot water for one minute.

Each serving equals:

HE: 1¼ Protein • 1 Bread • ¼ Fat •
19 Optional Calories

181 Calories • 1 gm Fat • 19 gm Protein •
24 gm Carbohydrate • 635 mg Sodium •
183 mg Calcium • 2 gm Fiber

DIABETIC: 2 Meat • 1½ Starch

Tuna Noodle and Celery Casserole

Tuna noodle casserole is a true American classic, and as I've traveled across the United States, I've been introduced to many variations on this beloved theme. Here's one with some extra-creamy flavor and just a bit of crunch.   ☻   Serves 4

> 1 (6-ounce) can white tuna, packed in water, drained and flaked
> ½ cup frozen peas, thawed
> 1 cup finely chopped celery
> 1 (10 ¾-ounce) can Healthy Request Cream of Celery Soup
> ⅓ cup Carnation Nonfat Dry Milk Powder
> ¼ cup water
> ¼ teaspoon black pepper
> 1 teaspoon dried parsley flakes
> 1½ cups hot cooked noodles, rinsed and drained

Preheat oven to 350 degrees. Spray an 8-by-8-inch baking dish with butter-flavored cooking spray. In a large bowl, combine tuna, peas, and celery. In a small bowl, combine celery soup, dry milk powder, and water. Stir in black pepper and parsley flakes. Add soup mixture to tuna mixture. Mix well to combine. Stir in noodles. Pour mixture into prepared baking dish. Bake for 40 to 45 minutes. Place baking dish on a wire rack and let set 5 minutes. Divide into 4 servings.

HINTS: 1. 1¼ cups uncooked noodles usually cooks to about 1½ cups.
   2. Thaw peas by placing in a colander and rinsing under hot water for one minute.

Each serving equals:

HE: 1 Bread • ¾ Protein • ½ Vegetable •
¼ Skim Milk • ½ Slider • 1 Optional Calorie

207 Calories • 3 gm Fat • 17 gm Protein •
28 gm Carbohydrate • 506 mg Sodium •
147 mg Calcium • 2 gm Fiber

DIABETIC: 2 Starch • 1½ Meat

Tuna Salad Pizza

If you're looking for some fun hot canapés for your next party, these creamy bites make a wonderful impression! Now that we can purchase prepared crescent rolls that are even lower in fat than the traditional kind, we can enjoy these special dishes often as part of a healthy lifestyle. ☻ Serves 8 (2 each)

> 1 (8-ounce) can Pillsbury Reduced Fat Crescent Rolls
> 1 (8-ounce) package Philadelphia fat-free cream cheese
> **1 recipe JO's Sour Cream**
> 1 tablespoon prepared horseradish sauce
> 2 (6-ounce) cans white tuna, packed in water,
>   drained and flaked
> 1/4 cup sliced green onion
> 1 teaspoon dried parsley flakes
> 1 3/4 cups finely chopped celery
> 2 cups shredded carrots

Preheat oven to 425 degrees. Spray a rimmed 9-by-13-inch cookie sheet with butter-flavored cooking spray. Pat rolls into pan being sure to seal perforations. Bake for 6 to 8 minutes or until crust is light golden brown. Place cookie sheet on a wire rack and allow to cool. Meanwhile, in a medium bowl, stir cream cheese with a spoon until soft. Blend in **JO's Sour Cream** and horseradish sauce. Add tuna, onion, and parsley flakes. Mix well to combine. Stir in celery and carrots. Spread mixture evenly over cooled crust. Refrigerate for at least 1 hour. Cut into 16 pieces.

Each serving equals:

HE: 1 1/4 Protein • 1 Bread • 1 Vegetable •
1/4 Skim Milk • 8 Optional Calories

205 Calories • 5 gm Fat • 19 gm Protein •
21 gm Carbohydrate • 600 mg Sodium •
95 mg Calcium • 1 gm Fiber

DIABETIC: 2 Meat • 1 Starch • 1 Fat • 1 Vegetable

Jiffy Shrimp Curry

Canned shrimp is a great recipe starter to keep in your pantry, especially if unexpected guests turn up for lunch. This dish delivers a lot of flavor with very little fuss, and tastes so creamy you'll have to work hard to convince any visiting dieters that they can eat and enjoy!   ☻   Serves 4 (full ½ cup)

1 (10¾-ounce) can Healthy Request Cream of
    Mushroom Soup
2 tablespoons dried onion flakes
½ teaspoon curry powder
1 recipe JO's Sour Cream
1 teaspoon cornstarch
1 (4.5-ounce drained weight) can tiny shrimp,
    rinsed and drained

In a large skillet sprayed with butter-flavored cooking spray, combine mushroom soup, onion flakes, and curry powder. Cook over medium heat for 5 minutes or until heated through, stirring often. In a small bowl, combine **JO's Sour Cream** and cornstarch. Stir into soup mixture. Add shrimp. Mix well to combine. Lower heat and simmer for 10 minutes, or until mixture is heated through, stirring often.

HINT: Good served over rice or fettuccine noodles.

Each serving equals:

HE: 1 Protein • ½ Skim Milk • ½ Slider •
4 Optional Calories

130 Calories • 2 gm Fat • 13 gm Protein •
15 gm Carbohydrate • 417 mg Sodium •
228 mg Calcium • 0 gm Fiber

DIABETIC: 1 Meat • ½ Skim Milk • ½ Starch
or 1 Meat • 1 Starch

Salmon Cakes with Creamed Pea Sauce

Finding new ways to add salmon to the menu is sure to help you get the dietary calcium you need to ensure strong and healthy bones. This delectable seafood dish is just as good as it is good for you—and the pea sauce offers a wonderful contrast in taste, texture, and color.   ☺   Serves 4

1 (14.5-ounce) can pink salmon, drained and flaked

14 small fat-free saltine crackers, made into fine crumbs

3 recipes JO's Evaporated Skim Milk☆

1 teaspoon dried onion flakes

2 teaspoons dried parsley flakes

3 tablespoons all-purpose flour

1 cup frozen peas, thawed

¼ teaspoon lemon pepper

In a large bowl, combine salmon, cracker crumbs, ¼ cup **JO's Evaporated Skim Milk**, onion flakes, and parsley flakes. Mix well to combine. Using a ⅓ cup measure as a guide, form mixture into 4 cakes. Place cakes in a large skillet sprayed with butter-flavored cooking spray. Brown cakes for about 5 minutes on each side. Meanwhile, in a covered jar, combine remaining 1¼ cups **JO's Evaporated Skim Milk** and flour. Shake well to blend. Pour mixture into a medium saucepan sprayed with butter-flavored cooking spray. Add peas and lemon pepper. Mix well to combine. Cook over medium heat for 5 minutes or until mixture thickens, stirring often. When serving, spoon about ⅓ cup pea sauce over each salmon cake.

HINT: Thaw peas by placing in a colander and rinsing under hot water for one minute.

Each serving equals:

HE: 3½ Protein • 1¼ Bread • ¾ Skim Milk

---

282 Calories • 6 gm Fat • 30 gm Protein •
27 gm Carbohydrate • 591 mg Sodium •
438 mg Calcium • 2 gm Fiber

DIABETIC: 3 Meat • 1 Starch • 1 Skim Milk

Creamy Italian
Baked Chicken Breasts

❄

There's something about the glorious aroma of Parmesan cheese drifting from your kitchen that says "Let's celebrate tonight!" This is a great dish to fix for family or friends on stressed-out nights, because once you slip it into the oven, you've got 45 minutes all to yourself! ♥ Serves 4

> 16 ounces skinned and boned uncooked chicken breasts,
>   cut into 4 pieces
> 1 (10¾-ounce) can Healthy Request Cream of Chicken Soup
> ¼ cup Kraft Fat Free Italian Dressing
> ⅔ cup Carnation Nonfat Dry Milk Powder
> ¼ cup (¾ ounce) grated Kraft fat-free Parmesan cheese
> 1 teaspoon dried parsley flakes

Preheat oven to 350 degrees. Spray an 8-by-8-inch baking dish with olive oil–flavored cooking spray. Evenly arrange chicken pieces in prepared baking dish. In a small bowl, combine chicken soup, Italian dressing, dry milk powder, Parmesan cheese, and parsley flakes. Evenly spoon soup mixture over chicken pieces. Cover and bake for 45 minutes. Uncover and continue to bake for 15 minutes or until chicken is tender. Divide into 4 servings. When serving, evenly spoon sauce over top of chicken pieces.

Each serving equals:

HE: 3¼ Protein • ½ Skim Milk • ½ Slider •
13 Optional Calories

247 Calories • 5 gm Fat • 35 gm Protein •
18 gm Carbohydrate • 681 mg Sodium •
157 mg Calcium • 1 gm Fiber

DIABETIC: 3 Meat • ½ Skim Milk • ½ Starch

Muffin Chicken Divan

Sure, you could have a plain old chicken sandwich for lunch or supper, but with just a little more effort, you can savor something special—and you deserve it! It's fun experimenting with the salsa to see just how hot you like it. You could even take this to work and put it in the microwave at lunchtime. **☻** Serves 4

2 English Muffins, split and toasted
1 full cup (6 ounces) diced cooked chicken breast
½ cup (one 2.5-ounce jar) sliced mushrooms, drained
1 cup chopped cooked broccoli
⅓ cup (1½ ounces) shredded Kraft reduced-fat
    Cheddar cheese
½ cup chunky salsa (mild, medium, or hot)
2 tablespoons Kraft fat-free mayonnaise

Preheat oven to 350 degrees. Spray a medium-sized cookie sheet with butter-flavored cooking spray. Evenly arrange muffin halves on prepared cookie sheet. In a medium bowl, combine chicken, mushrooms, broccoli, and Cheddar cheese. Add salsa and mayonnaise. Mix gently to combine. Spoon about ⅔ cup chicken mixture over each muffin half. Bake for 15 minutes. Serve at once.

Each serving equals:

HE: 2 Protein • 1 Vegetable • 1 Bread •
5 Optional Calories

188 Calories • 4 gm Fat • 20 gm Protein •
18 gm Carbohydrate • 514 mg Sodium •
179 mg Calcium • 2 gm Fiber

DIABETIC: 2 Meat • 1 Vegetable • 1 Starch

Chicken Roll-Ups

This is the kind of dish you'd expect to find only at a favorite Mexican restaurant, but now you can enjoy its delicious flavors at home. It's high on spicy, tangy taste—and pleasingly high in the calcium it delivers in every single bite! (The recipe I "revised" this from had 60 grams of fat per serving!) ☻ Serves 6

1 full cup (6 ounces) diced cooked chicken breast
1 (10¾-ounce) can Healthy Request Cream of Chicken Soup
½ cup chopped onion
1 teaspoon chili seasoning
1½ cups Yoplait plain fat-free yogurt☆
⅔ cup Carnation Nonfat Dry Milk Powder☆
2 teaspoons cornstarch☆
¾ cup (3 ounces) shredded Kraft reduced-fat
      Cheddar cheese☆
12 (6-inch) flour tortillas
¼ cup chopped green onion
⅓ cup (1½ ounces) sliced ripe olives
¾ cup water

Preheat oven to 350 degrees. Spray a 9-by-13-inch baking pan with butter-flavored cooking spray. In a large bowl, combine chicken, chicken soup, onion, and chili seasoning. Add ¾ cup yogurt, ⅓ cup dry milk powder, 1 teaspoon cornstarch, and ¼ cup Cheddar cheese. Mix well to combine. Spoon about 3 tablespoons mixture into each tortilla and roll up. Place filled tortillas in prepared baking pan seam side down. Lightly sprinkle onion, olives, and remaining ½ cup Cheddar cheese over top. In a small bowl, combine remaining ¾ cup yogurt, remaining ⅓ cup dry milk powder and remaining 1 teaspoon cornstarch. Add water. Mix well to combine. Spoon yogurt mixture evenly over top of roll-ups. Bake for 25 minutes. Place baking pan on a wire rack and let set for 5 minutes. Divide into 6 servings.

Each serving equals:

HE: 2 Bread • 1⅔ Protein • ⅔ Skim Milk •
¼ Vegetable • ¼ Fat • ¼ Slider •
13 Optional Calories

373 Calories • 9 gm Fat • 25 gm Protein •
48 gm Carbohydrate • 809 mg Sodium •
333 mg Calcium • 0 gm Fiber

DIABETIC: 2 Starch • 1½ Meat • 1 Skim Milk
or 3 Starch • 1½ Meat

Chicken Broccoli Skillet

My family is a fan of skillet suppers served up hot and creamy right at the table! This quick-fix delight combines favorite flavors of rice, chicken, and broccoli in a truly healthy and satisfyingly creamy dish. Just sit back and accept those compliments from every corner of the room! ◐ Serves 4

> 8 ounces skinned and boned uncooked chicken breast,
>   cut into 20 small pieces
> ½ cup chopped onion
> ½ cup (one 2.5-ounce jar) sliced mushrooms, drained
> 3 cups frozen cut broccoli, slightly thawed
> 1 (10¾-ounce) can Healthy Request Cream of
>   Chicken Soup
> ⅓ cup Carnation Nonfat Dry Milk Powder
> ¼ cup water
> 1 teaspoon dried parsley flakes
> ¼ teaspoon lemon pepper
> 2 cups hot cooked rice

In a large skillet sprayed with butter-flavored cooking spray, sauté chicken and onion for 5 minutes, stirring occasionally. Add mushrooms and broccoli. Mix well to combine. Continue cooking for 5 minutes, stirring occasionally. In a medium bowl, combine chicken soup, dry milk powder, water, parsley flakes, and lemon pepper. Add soup mixture to chicken mixture. Mix gently to combine. Lower heat, cover, and simmer for 10 minutes, stirring occasionally. For each serving, place ½ cup hot rice on a plate and spoon about 1 cup chicken mixture over top.

HINTS: 1. 1⅓ cups uncooked rice usually cooks to about 2
            cups.
        2. Thaw broccoli by placing in a colander and rinsing
            under hot water for two minutes.

Each serving equals:

HE: 2 Vegetable • 1½ Protein • 1 Bread •
¼ Skim Milk • ½ Slider • 5 Optional Calories

---

276 Calories • 4 gm Fat • 25 gm Protein •
34 gm Carbohydrate • 477 mg Sodium •
123 mg Calcium • 3 gm Fiber

DIABETIC: 2 Vegetable • 1½ Meat • 1½ Starch

Cozy Chicken and Green Bean Hot Dish

My philosophy is simple—I want it to be easy, healthy, and good. Here's a recipe that made Cliff smile as he gobbled it down! Okay, I know he'll devour anything with chicken and green beans, but I do enjoy making him happy, and pleasing myself too.

○ Serves 4

> ½ cup chopped onion
> 1 full cup (6 ounces) diced cooked chicken breast
> 4 cups (two 16-ounce cans) French style green beans, rinsed and drained
> **1 recipe JO's Dry Casserole Soup Mix**
> 1 cup water
> ¼ teaspoon black pepper
> 2 cups hot cooked noodles, rinsed and drained

In a large skillet sprayed with butter-flavored cooking spray, sauté onion and chicken for 5 minutes or until onion is tender. Stir in green beans. In a small bowl, combine **JO's Dry Casserole Soup Mix**, water, and black pepper. Add soup mixture to chicken mixture. Mix well to combine. Lower heat and simmer for 15 minutes, or until mixture is heated through, stirring often. For each serving, place ½ cup hot noodles on a plate and spoon about 1 cup chicken mixture over top.

HINTS: 1. 1¾ cups uncooked noodles usually cooks to about 2 cups.
    2. Also good served over rice or potatoes.

Each serving equals:

HE: 2¼ Vegetable • 1½ Protein • 1 Bread •
19 Optional Calories

---

247 Calories • 3 gm Fat • 20 gm Protein •
35 gm Carbohydrate • 366 mg Sodium •
121 mg Calcium • 3 gm Fiber

DIABETIC: 2 Vegetable • 1½ Meat • 1½ Starch

Chicken Rice Hash

You've heard of things that always go great together—love and marriage, horse and carriage, just like the song? Well, chicken and rice are terrific partners in culinary creations you'll prepare again and again. Don't skip the pimientos, even if you don't usually cook with them. They give this dish a pretty dash of color.

❂ Serves 4 (1 full cup)

1 (10¾-ounce) can Healthy Request Cream of Chicken Soup

1 cup skim milk

2 teaspoons reduced-sodium soy sauce

1½ cups (8 ounces) diced cooked chicken breast

½ cup frozen peas

¼ cup (one 2.5-ounce jar) chopped pimientos

¼ teaspoon black pepper

1 cup (3 ounces) uncooked instant rice

In a large skillet, combine chicken soup, skim milk, and soy sauce. Stir in chicken, peas, pimientos, and black pepper. Bring mixture to a boil. Stir in rice. Remove from heat, cover and let set for 5 minutes. Mix again just before serving.

Each serving equals:

HE: 2 Protein • 1 Bread • ¼ Skim Milk • ½ Slider •
1 Optional Calorie

207 Calories • 3 gm Fat • 23 gm Protein •
22 gm Carbohydrate • 548 mg Sodium •
93 mg Calcium • 1 gm Fiber

DIABETIC: 2 Meat • 1½ Starch

Stove-Top Hot Dish

Scrumptious. Luscious. The praise just keeps on coming when you stir up this impossibly creamy top-of-the-stove turkey feast. Here's perfect proof that stuffing is NOT just for the holidays, but healthy enough to belong on your plate throughout the year.

○ Serves 8

> 2 cups (one 16-ounce can) Healthy Request Chicken Broth
> 3 cups frozen carrot, broccoli, and cauliflower blend
> 2⅔ cups (6 ounces) Chicken-flavored Stove Top Stuffing Mix
> 1 (10¾-ounce) can Healthy Request Cream of Chicken Soup
> **1 recipe JO's Sour Cream**
> 1 teaspoon cornstarch
> 1½ cups (8 ounces) chopped cooked turkey breast
> 2 teaspoons dried parsley flakes

Preheat oven to 350 degrees. Spray a 9-by-13-inch baking pan with butter-flavored cooking spray. In a large saucepan, combine chicken broth and frozen vegetables. Bring mixture to a boil. Remove from heat. Stir in dry stuffing mix. In a small bowl, combine chicken soup, **JO's Sour Cream**, and cornstarch. Add soup mixture to stuffing mixture. Mix well to combine. Fold in turkey and parsley flakes. Spread mixture into prepared baking pan. Bake for 45 minutes. Place baking pan on a wire rack and let set for 5 minutes. Cut into 8 servings.

HINT: 1 cup frozen carrots, 1 cup frozen broccoli and 1 cup frozen cauliflower may be used in place of blended vegetables.

Each serving equals:

HE: 1 Bread • 1 Protein • ¾ Vegetable •
¼ Skim Milk • ¼ Slider • 4 Optional Calories

190 Calories • 2 gm Fat • 16 gm Protein •
27 gm Carbohydrate • 678 mg Sodium •
121 mg Calcium • 1 gm Fiber

DIABETIC: 1½ Starch • 1 Meat • 1 Vegetable

Wisconsin Turkey Sandwiches ❄

Every health-conscious cook needs great ideas for leftovers, and this is one of my favorite ways to "reinvent" the leftover turkey from Thanksgiving, or any festive family meal. This dish (inspired by the "state sandwich" of Wisconsin, according to a friend) was a mega-hit with Cliff and James!　　　**◐**　　Serves 6

> ½ cup finely chopped celery
> ½ cup finely chopped onion
> 1 (10¾-ounce) can Healthy Request Cream of Chicken Soup
> ⅔ cup Carnation Nonfat Dry Milk Powder
> ½ cup water
> 1 teaspoon dried parsley flakes
> 2 teaspoons ground sage
> ¼ teaspoon black pepper
> 2 full cups (12 ounces) diced cooked turkey breast
> 3 cups (4½ ounces) purchased dry bread cubes
> 6 reduced-calorie hamburger buns

In a large skillet sprayed with butter-flavored cooking spray, sauté celery and onion for 8 minutes or until tender. In a medium bowl, combine chicken soup, dry milk powder, water, parsley flakes, sage, and black pepper. Stir soup mixture into skillet. Add turkey and bread cubes. Mix well to combine. Continue cooking for 5 minutes or until mixture is heated through and bread becomes soft, stirring often. For each sandwich, spoon about ¾ cup turkey mixture between a bun.

HINT:　Brownberry unseasoned toasted bread cubes work great.

Each serving equals:

HE: 2 Protein • 2 Bread • ⅓ Vegetable •
⅓ Skim Milk • ¼ Slider • 10 Optional Calories

309 Calories • 5 gm Fat • 26 gm Protein •
40 gm Carbohydrate • 671 mg Sodium •
115 mg Calcium • 3 gm Fiber

DIABETIC: 2½ Starch • 2 Meat

Old-Fashioned Meat Loaf

All of us have happy memories of food, and we get a chance to relive good times in our lives when we enjoy those dishes again. But what if you've opted for a healthy lifestyle and your beloved recipes are just too high in fat to enjoy as often as you like? Healthy Exchanges to the rescue . . . with a great-tasting version of the ultimate old-time comfort food!   ♥   Serves 6

⅔ cup Carnation Nonfat Dry Milk Powder

⅔ cup water

16 ounces ground 90% lean ground turkey or beef

½ cup + 1 tablespoon (2¼ ounces) dried fine bread crumbs

2 tablespoons dried minced onion

¼ teaspoon black pepper

2 tablespoons Heinz Light Harvest or Healthy Choice Ketchup

Preheat oven to 350 degrees. Spray a 9-by-5-inch loaf pan with butter-flavored cooking spray. In a large bowl, combine dry milk powder and water. Add meat, bread crumbs, onion, and black pepper. Mix well to combine. Pat mixture into prepared loaf pan. Bake for 40 to 45 minutes. Evenly spoon ketchup over partially baked meat loaf. Continue baking for 10 minutes. Place loaf pan on a wire rack and let set for 5 minutes. Cut into 6 servings.

Each serving equals:

HE: 2 Protein • ⅓ Bread • ⅓ Skim Milk •
5 Optional Calories

183 Calories • 7 gm Fat • 17 gm Protein •
13 gm Carbohydrate • 194 mg Sodium •
118 mg Calcium • 1 gm Fiber

DIABETIC: 2 Meat • 1 Starch

Tom's Cheeseburger Meat Loaf ❄

Whenever you see "cheeseburger" in a recipe title, you can be sure I was thinking about my son Tommy when I stirred it up! Tom's a college grad now and working far from DeWitt, Iowa, but this is a meal I'll definitely cook for him when he returns for a visit. (This makes fantastic sandwiches the next day!) ☻ Serves 6

16 ounces ground 90% lean turkey or beef

6 tablespoons (1½ ounces) dried fine bread crumbs

½ cup dill pickle relish

¾ cup (3 ounces) shredded Kraft reduced-fat Cheddar cheese

½ cup chopped onion

1 cup (one 8-ounce can) Hunt's Tomato Sauce☆

1 tablespoon Brown Sugar Twin

Preheat oven to 350 degrees. Spray a 9-by-5-inch loaf pan with butter-flavored cooking spray. In a large bowl, combine meat, bread crumbs, pickle relish, Cheddar cheese, onion, and ⅓ cup tomato sauce. Mix well to combine. Pat mixture into prepared loaf pan. Stir Brown Sugar Twin into remaining ⅔ cup tomato sauce. Spread sauce mixture evenly over meat loaf. Bake for 45 to 55 minutes. Place loaf pan on a wire rack and let set for 5 minutes. Cut into 6 servings.

Each serving equals:

HE: 2⅔ Protein • 1 Vegetable • ⅓ Bread • 1 Optional Calorie

193 Calories • 9 gm Fat • 18 gm Protein • 10 gm Carbohydrate • 537 mg Sodium • 119 mg Calcium • 1 gm Fiber

DIABETIC: 2½ Meat • 1 Vegetable • ½ Starch

Salisbury Meat Loaf ❄

The basics of meat loaf are easy to remember, but it's the little extra touches that keep us from getting bored. I've substituted cornflakes for the usual bread crumbs here, and stirred up a special sauce that turns everyday meat loaf into one that's extra-tangy—and extra-good! ☻   Serves 6

1 full cup (2¼ ounces) cornflake crumbs
1½ cups (one 12-fluid-ounce can) Carnation Evaporated
*   Skim Milk☆*
16 ounces ground 90% lean turkey or beef
¼ teaspoon black pepper
1 teaspoon Worcestershire sauce
2 tablespoons dried onion flakes
2 tablespoons Heinz Light Harvest or Healthy Choice Ketchup
2 teaspoons dried parsley flakes
1 (10¾-ounce) can Healthy Request Cream of Mushroom Soup

Preheat oven to 350 degrees. Spray a 9-by-5-inch loaf pan with butter-flavored cooking spray. In a large bowl, soak cornflake crumbs in ½ cup evaporated skim milk for 2 to 3 minutes. Add meat, black pepper, Worcestershire sauce, onion flakes, and ketchup. Mix well to combine. Pat mixture into prepared loaf pan. Bake for 40 minutes. In a medium bowl, combine remaining 1 cup evaporated skim milk, parsley flakes, and mushroom soup. Spread soup mixture evenly over partially baked meat loaf. Continue baking for 10 to 15 minutes. Place loaf pan on a wire rack and let set for 5 minutes. Cut into 6 servings.

Each serving equals:

HE: 2 Protein • ½ Skim Milk • ½ Bread • ¼ Slider • 13 Optional Calories

219 Calories • 7 gm Fat • 19 gm Protein • 20 gm Carbohydrate • 400 mg Sodium • 244 mg Calcium • 1 gm Fiber

DIABETIC: 2 Meat • 1 Starch • ½ Skim Milk

Tom's Hamburger Milk Gravy Rice

Someday, I bet, I'll have enough recipes starring hamburger milk gravy to publish an entire cookbooklet! (And you can already guess who it'll be dedicated to!) You can hardly blame a mother for creating recipes to please her youngest child, can you? No matter how far he travels in his life, I know Tommy will always come running home to enjoy his favorites! ☕ Serves 4

> 8 ounces ground 90% lean turkey or beef
> 1½ cups (one 12-fluid-ounce can) Carnation Evaporated Skim Milk
> 3 tablespoons all-purpose flour
> ¼ teaspoon black pepper
> 1 teaspoon dried parsley flakes
> 1½ cups hot cooked rice
> ½ cup frozen peas
> 4 (¾ ounce) slices Kraft reduced-fat American cheese

Preheat oven to 350 degrees. Spray an 8-by-8-inch baking dish with butter-flavored cooking spray. In a medium skillet sprayed with butter-flavored cooking spray, brown meat. In a covered jar, combine evaporated skim milk and flour. Shake well to blend. Pour milk mixture into browned meat. Stir in black pepper and parsley flakes. Continue cooking for 5 minutes or until mixture thickens, stirring often. Add rice and peas. Mix gently to combine. Pour mixture into prepared baking dish. Place cheese slices evenly over top. Bake for 20 to 25 minutes or until mixture is bubbly. Place baking dish on a wire rack and let set for 5 minutes. Divide into 4 servings.

HINT: 1 cup uncooked rice usually cooks to about 1½ cups.

Each serving equals:

HE: 2½ Protein • 1¼ Bread • ¾ Skim Milk

300 Calories • 8 gm Fat • 24 gm Protein •
33 gm Carbohydrate • 501 mg Sodium •
412 mg Calcium • 1 gm Fiber

DIABETIC: 2 Meat • 1 Starch • 1 Skim Milk
or 2 Meat • 2 Starch

Hamburger Milk Gravy with Noodles

Just because you've got only a few minutes to prepare dinner, you don't have to settle for a frozen store-bought entree! This is so easy anyone can stir it up in just one pan and in under ten minutes. You get great taste, you get plenty of calcium, and you get to relax instead of spending hours in the kitchen.

○ Serves 4 (1 cup)

8 ounces ground 90% lean turkey or beef
1 recipe JO's Cream Sauce
1½ cups hot cooked noodles, rinsed and drained
½ cup frozen peas, thawed
½ cup (one 2.5-ounce jar) sliced mushrooms, drained

In a large skillet sprayed with butter-flavored cooking spray, brown meat. Stir in **JO's Cream Sauce**. Add noodles, peas, and mushrooms. Mix well to combine. Lower heat and simmer for 5 minutes or until mixture is heated through.

HINTS: 1. 1¼ cups uncooked noodles usually cooks to about 1½ cups.
   2. Thaw peas by placing in a colander and rinsing under hot water for one minute.

Each serving equals:

HE: 1½ Protein • 1¼ Bread • ¾ Skim Milk • ¼ Vegetable

266 Calories • 6 gm Fat • 21 gm Protein • 32 gm Carbohydrate • 235 mg Sodium • 226 mg Calcium • 2 gm Fiber

DIABETIC: 1½ Meat • 1½ Starch • ½ Skim Milk *or* 2 Starch • 1½ Meat

Hamburger Milk Gravy and Potatoes

I've promised from time to time to share some of my magic kitchen tricks with you, and this recipe illustrates one of my first and best: how a few tablespoons of flour combined with a can of evaporated skim milk transforms itself in minutes into thick, luscious gravy! *Abracadabra*—and you get to enjoy the results!

❍ Serves 4 (1 cup)

> 8 ounces ground 90% lean turkey or beef
> ½ cup chopped onion
> ½ teaspoon dried minced garlic
> 2 cups (12 ounces) chopped cooked potatoes
> ½ cup (one 2.5-ounce jar) sliced mushrooms, drained
> 1½ cups (one 12-fluid-ounce can) Carnation Evaporated
>   Skim Milk
> 3 tablespoons all-purpose flour
> 1 tablespoon dried parsley flakes

In a large skillet sprayed with butter-flavored cooking spray, brown meat, onion, and garlic. Add potatoes and mushrooms. Mix well to combine. Continue cooking for 5 to 7 minutes or until potatoes are browned, stirring occasionally. In a covered jar, combine evaporated skim milk and flour. Shake well to blend. Add milk mixture to meat mixture. Mix gently to combine. Stir in parsley flakes. Lower heat and simmer for 15 minutes or until mixture thickens, stirring often.

Each serving equals:

> HE: 1½ Protein • 1 Bread • ¾ Skim Milk •
> ½ Vegetable
>
> ---
> 285 Calories • 5 gm Fat • 22 gm Protein •
> 38 gm Carbohydrate • 282 mg Sodium •
> 330 mg Calcium • 3 gm Fiber
>
> ---
> DIABETIC: 1½ Meat • 1 Starch • 1 Skim Milk
> *or* 2 Starch • 1½ Meat

Simple Pleasures Skillet

Some of the recipes our families love most are the simplest and even the least expensive to prepare. This speedy skillet supper uses ingredients we keep on the pantry shelf, but when you stir them all together and spice it with love, you've got a winner every time.

○ Serves 6 (1 cup)

> 16 ounces ground 90% lean turkey or beef
> 2 cups (one 16-ounce can) cut green beans,
>   rinsed and drained
> 3 cups hot cooked noodles, rinsed and drained
> 1 (10¾-ounce) can Healthy Request Cream of
>   Mushroom Soup
> ⅔ cup Carnation Nonfat Dry Milk Powder
> ½ cup water
> ¼ teaspoon black pepper

In a large skillet sprayed with butter-flavored cooking spray, brown meat. Stir in green beans and noodles. In a small bowl, combine mushroom soup, dry milk powder, water, and black pepper. Add soup mixture to meat mixture. Mix well to combine. Lower heat, cover, and simmer for 10 minutes, stirring occasionally.

HINT: 2½ cups uncooked noodles usually cooks to about 3 cups.

Each serving equals:

HE: 2 Protein • 1 Bread • ⅔ Vegetable •
⅓ Skim Milk • ¼ Slider • 8 Optional Calories

---

276 Calories • 8 gm Fat • 21 gm Protein •
30 gm Carbohydrate • 319 mg Sodium •
147 mg Calcium • 2 gm Fiber

DIABETIC: 2 Meat • 1½ Starch • 1 Vegetable

Pork à la King

My theory about this dish is that the "king," whoever he was, got tired of having chicken à la king—and wanted a change! Here's my suggestion, perfect for using up leftover pork roast in a simply delicious way. 🌑 Serves 4 (1 full cup)

1 recipe JO's Cream Sauce

½ *cup (one 2.5-ounce jar) sliced mushrooms, drained*

¼ *cup (one 2.5-ounce jar) diced pimientos*

1 *cup frozen peas, thawed*

¼ *teaspoon thyme*

1½ *cups (8 ounces) diced cooked lean roast pork*

In a medium saucepan, combine warm **JO's Cream Sauce**, mushrooms, pimientos, peas, and thyme. Gently stir in pork. Continue cooking for 5 minutes or until mixture is heated through, stirring often.

HINTS: 1. Good served over noodles, rice, or toast.
2. Thaw peas by placing in a colander and rinsing under hot water for one minute.

Each serving equals:

HE: 2 Protein • ¾ Bread • ¾ Skim Milk • ¼ Vegetable

191 Calories • 3 gm Fat • 21 gm Protein • 20 gm Carbohydrate • 217 mg Sodium • 237 mg Calcium • 3 gm Fiber

DIABETIC: 2 Meat • 1 Starch • ½ Skim Milk

Creamy Macaroni Taco Casserole

Did you know that in addition to tickling your taste buds and occasionally igniting your tongue, spicy foods help you burn a few extra calories? It's true! That's as good a reason as any for including lots of Mexican-inspired dishes in your menu planning. Here's one that emphasizes intense flavors more than heat!

☾ Serves 6

> 8 ounces ground 90% lean turkey or beef
> 1 (10¾-ounce) can Healthy Request Cream of
>   Mushroom Soup
> **2 recipes JO's Evaporated Skim Milk**
> 2 teaspoons taco seasoning
> 3 cups hot cooked elbow macaroni, rinsed and drained
> ¾ cup (3 ounces) shredded Kraft reduced-fat Cheddar cheese
> 2 cups finely shredded lettuce
> 1 cup finely chopped fresh tomatoes
> 6 tablespoons Land O Lakes no-fat sour cream

Preheat oven to 350 degrees. Spray an 8-by-8-inch baking dish with olive oil–flavored cooking spray. In a large skillet sprayed with olive oil–flavored cooking spray, brown meat. In a small bowl, combine mushroom soup, **JO's Evaporated Skim Milk**, and taco seasoning. Add soup mixture to browned meat. Mix well to combine. Stir in macaroni and Cheddar cheese. Pour mixture into prepared baking dish. Bake for 25 to 30 minutes. Place baking dish on a wire rack and let set for 5 minutes. Cut into 6 servings. When serving, top each piece with ⅓ cup shredded lettuce, about 2 tablespoons chopped tomatoes, and 1 tablespoon sour cream.

HINT: 2 cups uncooked macaroni usually cooks to about
        3 cups.

Each serving equals:

HE: 1⅔ Protein • 1 Bread • 1 Vegetable •
⅓ Skim Milk • ½ Slider • 3 Optional Calories

267 Calories • 7 gm Fat • 18 gm Protein •
33 gm Carbohydrate • 432 mg Sodium •
253 mg Calcium • 1 gm Fiber

DIABETIC: 2 Starch • 1½ Meat

Salisbury "Steak" and Rice with Vegetables

This baked version of that popular classic, Salisbury steak, starts life in a skillet but develops its rich flavors as a result of the tangy sauce that surrounds it during an hour in your oven. My son-in-law John told me it tastes even better than it looks—and it looks absolutely delectable! ☉ Serves 4

> 2 cups cold cooked rice
> 1 cup (one 8-ounce can) sliced carrots, rinsed and drained
> 1 cup (one 8-ounce can) cut green beans, rinsed and drained
> 1 (10¾-ounce) can Healthy Request Cream of Mushroom Soup☆
> 8 ounces ground 90% lean turkey or beef
> 7 small fat-free saltine crackers, made into fine crumbs
> ½ cup chopped onion or 2 tablespoons dried onion flakes
> ¾ cup water☆
> ⅛ teaspoon black pepper
> 1 tablespoon Worcestershire sauce
> 3 tablespoons Heinz Lite Harvest or Healthy Choice Ketchup
> ⅓ cup Carnation Nonfat Dry Milk Powder

Preheat oven to 350 degrees. Spray an 8-by-8-inch baking dish with butter-flavored cooking spray. In a large bowl, combine rice, carrots, green beans, and ⅓ cup mushroom soup. Spread mixture into prepared baking dish. In a medium bowl, combine meat, cracker crumbs, onion, ¼ cup water, and black pepper. Form mixture into 4 patties. Place patties in a large skillet sprayed with butter-flavored cooking spray and brown for 3 to 4 minutes on each side. Arrange browned patties over rice mixture. In a medium bowl, combine remaining mushroom soup, Worcestershire sauce, ketchup, dry milk powder, and remaining ½ cup water. Spread soup mixture evenly over top. Cover and bake for 45 minutes. Uncover and continue baking for 10 to 15 minutes. Place baking dish on a wire rack and let set for 5 minutes. Divide into 4 servings.

HINT: 1⅓ cups uncooked rice usually cooks to about 2 cups.

Each serving equals:

HE: 1½ Protein • 1¼ Vegetable • 1 Bread •
¼ Skim Milk • ½ Slider • 13 Optional Calories

267 Calories • 7 gm Fat • 15 gm Protein •
36 gm Carbohydrate • 620 mg Sodium •
139 mg Calcium • 1 gm Fiber

DIABETIC: 2 Meat • 1½ Starch • 1 Vegetable

Pam's Lasagna

My daughter-in-law Pam (along with my son James and those irresistible grandbabies of mine) loves Italian food, so this recipe was conceived with her smile in mind! Lasagna may seem like a complicated dish best left to restaurants, but this microwaved "masterpiece" will surprise you with its ease and scrumptious-looking result.   ◐   Serves 8

8 ounces ground 90% lean turkey or beef

¼ cup chopped onion

1¾ cups (one 14½-ounce can) stewed tomatoes, coarsely chopped and undrained

1¾ cups (one 15-ounce can) Hunt's Chunky Tomato Sauce

½ cup (one 2.5-ounce jar) sliced mushrooms

1½ teaspoons Italian seasoning

2 cups fat-free cottage cheese

¼ cup (¾ ounce) grated Kraft fat-free Parmesan cheese

1 egg, slightly beaten, or equivalent in egg substitute

1 tablespoon dried parsley flakes

6 uncooked lasagna noodles

1½ cups (6 ounces) shredded Kraft reduced-fat mozzarella cheese

Place meat and onion in a plastic colander and set colander in a glass pie plate. Cover and microwave on HIGH (100% power) for 5 minutes or until meat is browned, stirring after 2 minutes. In a medium bowl, combine browned meat mixture, undrained stewed tomatoes, tomato sauce, mushrooms, and Italian seasoning. In another medium bowl, combine cottage cheese, Parmesan cheese, egg, and parsley flakes. Spray a 9-by-13-inch baking dish with olive oil–flavored cooking spray. In prepared baking dish, layer 1 cup meat sauce, 3 uncooked noodles, ½ of cottage cheese mixture, and ¾ cup mozzarella cheese. Repeat layers, starting with 1 cup meat sauce. Spread remaining meat mixture over top. Cover and microwave on HIGH for 15 minutes. Turn baking dish and continue microwaving on MEDIUM (50% power) for 18 to 20 minutes.

Uncover and place baking dish on a wire rack and let set for 5 minutes. Divide into 8 servings.

Each serving equals:

HE: 2½ Protein • 1½ Vegetable • 1 Bread

186 Calories • 6 gm Fat • 18 gm Protein •
15 gm Carbohydrate • 727 mg Sodium •
183 mg Calcium • 2 gm Fiber

DIABETIC: 2½ Meat • 1 Vegetable • 1 Starch

Pam's Birthday Shepherd's Casserole

❋

Does it seem as if we're always celebrating something out here at Healthy Exchanges? Besides the regular birthday parties for our staff each month, I do enjoy creating special dishes to honor the favorite flavors of the people in my life. This new wrinkle on classic shepherd's pie was one of my gifts to Pam this year!

☻ Serves 4

8 ounces 90% lean turkey or beef
1 (10¾-ounce) can Healthy Request Tomato Soup
2 cups (one 16-ounce can) cut green beans,
     rinsed and drained
2 cups (one 16-ounce can) sliced carrots,
     rinsed and drained
1¾ cups water
1⅓ cups (3 ounces) instant potato flakes
⅓ cup Carnation Nonfat Dry Milk Powder
1 teaspoon dried parsley flakes
¼ teaspoon black pepper
½ cup (one 2.5-ounce jar) sliced mushrooms, drained

Preheat oven to 350 degrees. Spray an 8-by-8-inch baking dish with butter-flavored cooking spray. In a large skillet sprayed with butter-flavored cooking spray, brown meat. Stir in tomato soup, green beans, and carrots. Evenly spoon mixture into prepared baking dish. In a medium saucepan, bring water to a boil. Remove from heat. Stir in potato flakes and dry milk powder. Add parsley flakes, black pepper, and mushrooms. Mix well to combine. Spread potato mixture evenly over meat mixture. Bake for 25 to 30 minutes. Place baking dish on a wire rack and let set for 5 minutes. Divide into 4 servings.

Each serving equals:

HE: 2¼ Vegetable • 1½ Protein • 1 Bread •
¼ Skim Milk • ½ Slider • 5 Optional Calories

246 Calories • 6 gm Fat • 16 gm Protein •
32 gm Carbohydrate • 446 mg Sodium •
121 mg Calcium • 4 gm Fiber

DIABETIC: 2 Vegetable • 1½ Meat • 1½ Starch

Cornbread Pie

My very first healthy recipe was a main-dish pie, and they've been important staples of my cookbooks since the beginning! This meaty, cheesy delight bakes the special goodness of cornbread right into the recipe, blending all those varied flavors into a dish fit for a king—or my husband! ☻ Serves 6

> 8 ounces ground 90% lean turkey or beef
> ½ cup chopped onion
> ½ cup chopped green bell pepper
> 1 cup frozen whole kernel corn, thawed
> 1 (10¾-ounce) can Healthy Request Tomato Soup
> 1 teaspoon chili seasoning
> ¾ cup (4.5 ounces) yellow cornmeal
> 1 tablespoon all-purpose flour
> 1 tablespoon Sugar Twin or Sprinkle Sweet
> 1½ teaspoons baking powder
> ⅔ cup Carnation Nonfat Dry Milk Powder
> ¾ cup water
> 1 egg or equivalent in egg substitute
> ¼ cup (¾ ounce) grated Kraft fat-free Parmesan cheese
> 1 teaspoon dried parsley flakes

Preheat oven to 400 degrees. Spray a 10-inch deep-dish pie plate with olive oil–flavored cooking spray. In a large skillet sprayed with olive oil–flavored cooking spray, brown meat, onion, and green pepper. Stir in corn, tomato soup, and chili seasoning. Pour mixture into prepared pie plate. In a medium bowl, combine cornmeal, flour, Sugar Twin, and baking powder. In a small bowl, combine dry milk powder and water. Stir in egg. Add milk mixture to cornmeal mixture. Mix gently to combine. Fold in Parmesan cheese and parsley flakes. Spoon cornmeal mixture evenly over meat mixture. Bake for 20 to 25 minutes. Place pie plate on a wire rack and let set for 5 minutes. Cut into 6 servings.

HINT: Thaw corn by placing in a colander and rinsing under hot
water for one minute.

Each serving equals:

HE: 1⅓ Protein • 1⅓ Bread • ⅓ Vegetable •
⅓ Skim Milk • ¼ Slider • 17 Optional Calories

212 Calories • 5 gm Fat • 14 gm Protein •
34 gm Carbohydrate • 418 mg Sodium •
175 mg Calcium • 3 gm Fiber

DIABETIC: 2 Starch • 1½ Meat

Rio Grande Pizza Strata ❄

If you're not familiar with the word *strata*, it describes a wonderfully good, Italian-inspired baked dish that stirs together eggs, bread, cheese, meat, and tomato sauce into a true party-on-a-plate! This recipe blends in a bit of Southern spiciness to make the finished product something to cheer! ☻ Serves 6

8 ounces 90% lean turkey or beef
2 teaspoons chili seasoning
2 cups chunky salsa (mild, medium, or hot)
1 cup (one 8-ounce can) Hunt's Tomato Sauce
8 slices reduced-calorie white bread
¾ cup (3 ounces) shredded Kraft reduced-fat
      Cheddar cheese☆
2 eggs or equivalent in egg substitute
2 cups skim milk
1 teaspoon dried parsley flakes

Preheat oven to 350 degrees. Spray a 9-by-9-inch cake pan with olive oil–flavored cooking spray. In a large skillet sprayed with olive oil–flavored cooking spray, brown meat. Stir in chili seasoning, salsa, and tomato sauce. Lower heat and simmer for 5 minutes. Remove from heat. Place 4 slices of bread in prepared cake pan. Sprinkle half of Cheddar cheese over bread. Spoon half of meat mixture over cheese. Repeat layers. In a medium bowl, beat eggs with a wire whisk until frothy. Add skim milk and parsley flakes. Mix well to combine. Pour milk mixture evenly over top. Bake for 60 minutes or until edges are lightly browned and center is firm. Place cake pan on a wire rack and let set for 5 minutes. Divide into 6 servings.

HINTS: 1. If you can find Wonder Sourdough Fat Free Bread, it works great.
2. Strata can be covered and refrigerated up to 24 hours before baking.

Each serving equals:

HE: 2 Protein (⅓ limited) • 1⅓ Vegetable • ⅔ Bread • ⅓ Skim Milk

232 Calories • 8 gm Fat • 19 gm Protein • 22 gm Carbohydrate • 811 mg Sodium • 339 mg Calcium • 1 gm Fiber

DIABETIC: 1½ Meat • 1½ Vegetable • ½ Starch

"Sausage" Supper Skillet

I wasn't exactly trying to fool your taste buds when I named this dish, but your family will insist that there's got to be sausage in this tasty blend! This low-fat but highly flavored skillet dish gives you the satisfaction of eating what you love, but without the calories and fat you don't want anymore.

☯ Serves 4 (1 full cup)

8 ounces ground 90% lean turkey or beef
½ teaspoon poultry seasoning
¼ teaspoon ground sage
¼ teaspoon garlic powder
½ cup chopped onion
½ cup (one 2.5-ounce jar) sliced mushrooms
3 cups shredded cabbage
⅓ cup JO's Dry Casserole Soup Mix
1 cup water
2 cups hot cooked noodles, rinsed and drained

In a large skillet sprayed with butter-flavored cooking spray, brown meat. Stir in poultry seasoning, sage, garlic powder, onion, mushrooms, and cabbage. Continue cooking for 8 to 10 minutes or until cabbage is tender and lightly browned, stirring often. In a small bowl, combine **JO's Dry Casserole Soup Mix** and water. Add to meat mixture. Mix well to combine. Stir in noodles. Lower heat and simmer for 5 minutes or until mixture is heated through, stirring occasionally.

HINTS: 1. 1¾ cups uncooked noodles usually cooks to about 2 cups.
2. Purchased coleslaw mix may be used in place of shredded cabbage.

Each serving equals:

HE: 2 Vegetable • 1½ Protein • 1 Bread •
19 Optional Calories

234 Calories • 6 gm Fat • 16 gm Protein •
29 gm Carbohydrate • 301 mg Sodium •
171 mg Calcium • 3 gm Fiber

DIABETIC: 1½ Meat • 1½ Starch • 1 Vegetable

"Sausage" Quiche Squares

For a family brunch or weekend lunch when you're expecting a crowd, this is a terrific way to win their hearts and please their tummies! You'll be as dazzled as they are when you realize how simple it is to prepare—and how it'll tempt the fussiest eater.

● Serves 12 (2 pieces each)

2 tablespoons Kraft Fat Free Italian Dressing

8 ounces ground 90% lean turkey or beef

1/2 teaspoon poultry seasoning

1/4 teaspoon ground sage

1/4 teaspoon garlic powder

1 (8-ounce) can Pillsbury Reduced Fat Crescent Rolls

3/4 cup (3 ounces) shredded Kraft reduced-fat Cheddar cheese

3/4 cup (3 ounces) shredded Kraft reduced-fat
    mozzarella cheese

1 1/2 cups Yoplait plain fat-free yogurt

2/3 cup Carnation Nonfat Dry Milk Powder

2 eggs or equivalent in egg substitute

1 (10 3/4-ounce) can Healthy Request Cream of Chicken Soup

1/2 cup finely chopped onion

2 teaspoons Worcestershire sauce

1 teaspoon Italian seasoning

1/2 teaspoon lemon pepper

2 tablespoons dried parsley flakes

Preheat oven to 350 degrees. Spray a rimmed 10-by-15-inch cookie sheet with olive oil–flavored cooking spray. Pour Italian dressing into a large skillet. Add meat, poultry seasoning, sage, and garlic powder. Mix well to combine. Brown meat mixture. Place skillet on a wire rack and allow to cool. Meanwhile, pat rolls into prepared cookie sheet to form a crust, being sure to seal perforations. Evenly sprinkle Cheddar cheese, mozzarella cheese, and browned meat mixture over rolls. In a blender container, combine yogurt, dry milk powder, eggs, chicken soup, onion, Worcestershire sauce, Italian seasoning, lemon pepper, and parsley flakes. Cover and process on HIGH for 30 seconds or until mixture is smooth. Pour mixture evenly over top, spreading with a spatula as necessary to cover entire crust. Bake for 40 to 45 minutes or until golden brown. Place cookie sheet on a wire rack and allow to cool for 30 minutes. Cut into 24 pieces. Good warm or cold.

HINT: Do not use inexpensive rolls, as they don't cover the pan properly.

Each serving equals:

HE: 1⅓ Protein • ⅔ Bread • ⅓ Skim Milk • 16 Optional Calories

217 Calories • 9 gm Fat • 17 gm Protein • 17 gm Carbohydrate • 551 mg Sodium • 196 mg Calcium • 0 gm Fiber

DIABETIC: 1 Protein • 1 Starch • 1 Fat

Creamed Pork Casserole

Many people expect a healthy cookbook to eliminate all the meats they've grown to love, but now that we can buy delectably lean versions of everything from beef to pork, we no longer need to feel deprived of our favorites! This baked dish won four stars from Cliff, who's always liked pork, especially when I serve it with string beans. ☻ Serves 4

> 4 (4-ounce) lean tenderized pork tenderloins
> **1 recipe JO's Cream Sauce**
> ½ cup (one 2.5-ounce jar) sliced mushrooms, drained
> 2 cups (one 16-ounce can) cut green beans, rinsed and drained
> ¼ teaspoon lemon pepper
> 3 tablespoons (¾ ounce) dried fine bread crumbs

Preheat oven to 375 degrees. Spray an 8-by-8-inch baking dish with butter-flavored cooking spray. Lightly brown tenderloins in a large skillet sprayed with butter-flavored cooking spray. Meanwhile, in a medium saucepan sprayed with butter-flavored cooking spray, combine warm **JO's Cream Sauce**, mushrooms, green beans, and lemon pepper. Mix gently to combine. Place browned meat in prepared baking dish. Pour sauce mixture evenly over meat. Evenly sprinkle bread crumbs over top. Bake for 45 minutes. Place baking dish on a wire rack and let set for 5 minutes. Divide into 4 servings.

Each serving equals:

HE: 3 Protein • 1¼ Vegetable • ¾ Skim Milk • ½ Bread

---

270 Calories • 6 gm Fat • 33 gm Protein • 21 gm Carbohydrate • 294 mg Sodium • 266 mg Calcium • 2 gm Fiber

---

DIABETIC: 3 Meat • 1 Vegetable • 1 Starch

Heartland Pork Stroganoff

Classic stroganoff is more purely creamy than this recipe, which stirs in the rich sweetness of tomato soup. But considering the affection many Russians have for things "red," I bet they'd agree this scrumptious pork is a winner! ☻ Serves 4 (1 cup)

1 cup chopped onion
½ cup (one 2.5-ounce jar) sliced mushrooms, drained
1 full cup (6 ounces) diced lean cooked roast pork
2 cups cooked noodles, rinsed and drained
1 (10¾-ounce) can Healthy Request Tomato Soup
⅓ cup Carnation Nonfat Dry Milk Powder
¼ cup water
¼ cup Land O Lakes no-fat sour cream
1 teaspoon dried parsley flakes
¼ teaspoon black pepper

In a large skillet sprayed with butter-flavored cooking spray, sauté onion for about 5 minutes or until tender. Add mushrooms, pork, and noodles. Mix well to combine. In a small bowl, combine tomato soup, dry milk powder, and water. Stir in sour cream, parsley flakes, and black pepper. Add soup mixture to pork mixture. Mix well to combine. Lower heat and simmer for 15 minutes, or until mixture is heated through, stirring often.

HINTS:　1. 1¾ cups uncooked noodles usually cooks to about 2 cups.
　　　　2. Lean beef minute steaks may be used in place of pork.

Each serving equals:

HE: 1½ Protein • 1 Bread • ¾ Vegetable • ¼ Skim Milk • ¾ Slider

277 Calories • 5 gm Fat • 20 gm Protein • 38 gm Carbohydrate • 408 mg Sodium • 125 mg Calcium • 3 gm Fiber

DIABETIC: 2 Starch • 1½ Meat • 1 Vegetable

Potato Dumplings with Roast Pork and Sauerkraut

❄

Cold weather seems to call for hearty dishes that fill the kitchen with a cozy aroma of Old World flavors! My Bohemian heritage is showing in this dish. My mother and grandmother made the best potato dumplings with pork in the whole world. I wanted the flavor of theirs without the excess fats. I think they'd agree I succeeded, even though I didn't use "traditional" ingredients.

◐   Serves 4

> 1½ cups (8 ounces) diced lean cooked roast pork
> 3½ cups (two 14.5-ounce cans) Bavarian-style sauerkraut,
>   undrained
> ⅔ cup (1½ ounces) instant potato flakes
> ¾ cup all-purpose flour
> 1½ teaspoons baking powder
> 2 tablespoons fresh chopped parsley or 2 teaspoons
>   dried parsley flakes
> **1 recipe JO's Evaporated Skim Milk**
> 1 egg or equivalent in egg substitute

In a large saucepan, combine pork and sauerkraut. Cook over medium heat for 10 minutes, stirring occasionally. Meanwhile in a large bowl, combine potato flakes, flour, baking powder, parsley, **JO's Evaporated Skim Milk,** water, and egg. Mix well to moisten. Let set for 3 minutes. Drop by scant ¼-cup measure into hot sauerkraut mixture to form 4 dumplings. Cover tightly, lower heat, and simmer for 10 to 15 minutes, or until dumplings are tender. Divide into 4 servings.

HINT: If you can't find Bavarian-style sauerkraut, use regular sauerkraut, ½ teaspoon caraway seeds, and 1 teaspoon Brown Sugar Twin.

Each serving equals:

HE: 2¼ Protein (¼ limited) • 1¾ Vegetable • 1½ Bread • ¼ Skim Milk

---

281 Calories • 5 gm Fat • 21 gm Protein • 38 gm Carbohydrate • 1236 mg Sodium • 269 mg Calcium • 7 gm Fiber

---

DIABETIC: 2 Meat • 2 Vegetable • 1½ Starch

Asparagus Ham Lasagna Casserole

I bet your family will be as surprised and pleased as mine was to bite into a lasagna dish—and discover some tasty changes! Instead of the classic filling, I've substituted chunks of ham and asparagus, which give new interest to this popular recipe. Of course, I'd never eliminate those luscious cheeses. . . . ● Serves 4

> *3 recipes JO's Evaporated Skim Milk*
> *3 tablespoons all-purpose flour*
> *¼ cup finely chopped onion*
> *½ teaspoon dried minced garlic*
> *1 teaspoon Italian seasoning*
> *2 cups cooked mini lasagna noodles, rinsed and drained*
> *1 full cup (6 ounces) diced Dubuque 97% fat-free ham or any*
>   *extra-lean ham*
> *1¾ cups (one 14.5-ounce can) cut asparagus, rinsed and*
>   *drained*
> *¾ cup (3 ounces) shredded Kraft reduced-fat mozzarella cheese*
> *¼ cup (¾ ounce) grated Kraft fat-free Parmesan cheese*

Spray an 8-by-8-inch glass baking dish with butter-flavored cooking spray. In a covered jar, combine **JO's Evaporated Skim Milk** and flour. Cover and shake well to blend. Pour mixture into an 8-cup glass measuring bowl. Add onion, garlic, and Italian seasoning. Mix well to combine. Cover with waxed paper. Microwave on HIGH (100% power) for 3 minutes, stirring after every minute. Spread ⅓ cup of the sauce in bottom of prepared baking dish. Layer half of noodles over sauce. Sprinkle half of the ham and asparagus over noodles. Top with half of mozzarella cheese and half of remaining sauce. Repeat layers. Cover and microwave on BAKE (60% power) for 6 minutes, turning dish after 3 minutes. Sprinkle Parmesan cheese evenly over top. Continue microwaving, uncovered, on BAKE for an additional 3 minutes. Place baking dish on a wire rack and let set for 5 minutes. Divide into 4 servings.

HINTS: 1. 1¾ cups uncooked mini lasagna noodles usually cooks to about 2 cups.
2. If you can't find mini lasagna noodles, use regular medium-width noodles.

Each serving equals:

HE: 2¼ Protein • 1¼ Bread • 1 Vegetable • ¾ Skim Milk

327 Calories • 7 gm Fat • 26 gm Protein • 40 gm Carbohydrate • 950 mg Sodium • 376 mg Calcium • 4 gm Fiber

DIABETIC: 2½ Meat • 1 Starch • 1 Vegetable • 1 Skim Milk

Cliff's Special Macaroni and Ham

❄

This easy dinner is a kind of stove-top macaroni-and-cheese dish, but it's anything but ordinary! The blend of Parmesan and Swiss cheeses gives it a special tanginess, and every little bite of ham gives it a real hearty kick.   ☻   Serves 6

> ½ cup chopped onion
> 1 full cup (6 ounces) diced Dubuque 97% fat-free ham or any extra-lean ham
> ½ cup (one 2.5-ounce jar) sliced mushrooms, drained
> 3 cups cooked elbow macaroni, rinsed and drained
> 1 (10¾-ounce) can Healthy Request Cream of Mushroom Soup
> ⅔ cup Carnation Nonfat Dry Milk Powder
> ½ cup water
> 1 teaspoon dried parsley flakes
> ¼ cup (¾ ounce) grated Kraft fat-free Parmesan cheese
> 3 (¾-ounce) slices Kraft reduced-fat Swiss cheese

In a large skillet sprayed with butter-flavored cooking spray, sauté onion and ham for 5 minutes or until onion is tender. Stir in mushrooms and macaroni. In a small bowl, combine mushroom soup, dry milk powder, water, parsley flakes, and Parmesan cheese. Add soup mixture to ham mixture. Mix well to combine. Cut Swiss cheese slices in half. Evenly space cheese pieces over macaroni mixture. Lower heat, cover, and simmer for 2 to 3 minutes or until cheese melts. Divide into 6 servings.

HINT: 2 cups uncooked macaroni usually cooks to about 3 cups.

Each serving equals:

HE: 1⅓ Protein • 1 Bread • ⅓ Vegetable •
⅓ Skim Milk • ¼ Slider • 8 Optional Calories

---

207 Calories • 3 gm Fat • 13 gm Protein •
32 gm Carbohydrate • 600 mg Sodium •
169 mg Calcium • 2 gm Fiber

DIABETIC: 2 Starch • 1½ Meat

Cabbage Patch Skillet

I love the convenience of purchased coleslaw mix—it's such a perfect dinner starter for busy people! Here, I've stirred in ham and corn, then made it all hold hands with a creamy thick gravy that's rich without being "bad" for you. ☻ Serves 4 (1 full cup)

> 4 cups purchased coleslaw mix
> 1 full cup (6 ounces) diced Dubuque 97% fat-free ham or
>   any extra-lean ham
> 1½ cups frozen corn, thawed
> 1½ cups (one 12-fluid-ounce can) Carnation Evaporated
>   Skim Milk
> 3 tablespoons all-purpose flour
> ¼ teaspoon black pepper
> 1 teaspoon dried parsley flakes

In a large skillet sprayed with butter-flavored cooking spray, sauté coleslaw mix and ham for 10 minutes or until cabbage is tender, stirring often. Add corn. Mix well to combine. In a covered jar, combine evaporated skim milk and flour. Shake well to blend. Pour milk mixture into cabbage mixture. Stir in black pepper and parsley flakes. Lower heat and simmer for 5 minutes or until mixture thickens, stirring often.

HINTS: 1. 3 cups shredded cabbage and 1 cup shredded carrots may be used in place of purchased coleslaw mix.
2. Thaw corn by placing in a colander and rinsing under hot water for one minute.

Each serving equals:

HE: 2 Vegetable • 1 Protein • 1 Bread • ¾ Skim Milk

222 Calories • 2 gm Fat • 17 gm Protein •
34 gm Carbohydrate • 493 mg Sodium •
314 mg Calcium • 3 gm Fiber

DIABETIC: 1 Vegetable • 1 Meat • 1 Starch •
1 Skim Milk

Real Man Quiche

Remember that humor book a few years ago that suggested eating quiche might not be "manly"? Well, thank heavens that real men don't decide what to eat by reading other people's opinions—they vote with their mouths! This ham-and-cheese pie will definitely please the men in your life!   ♥   Serves 6

> *1½ cups (one 12-fluid-ounce can) Carnation Evaporated*
>   *Skim Milk*
> *½ cup + 1 tablespoon Bisquick Reduced Fat Baking Mix*
> *3 eggs or equivalent in egg substitute*
> *½ cup frozen peas, thawed*
> *1 full cup (6 ounces) diced Dubuque 97% fat-free ham*
>   *or any extra-lean ham*
> *½ cup (one 2.5-ounce jar) sliced mushrooms, drained*
> *¼ cup chopped onion*
> *¼ teaspoon lemon pepper*
> *Full ½ cup (2¼ ounces) shredded Kraft reduced-fat*
>   *Cheddar cheese*

Preheat oven to 350 degrees. Spray a deep-dish 10-inch pie plate with butter-flavored cooking spray. In a large bowl, combine evaporated skim milk, baking mix, and eggs. Add peas, ham, mushrooms, onion, and lemon pepper. Mix well to combine. Pour mixture into prepared pie plate. Evenly sprinkle Cheddar cheese over top. Bake for 35 to 45 minutes or until center is set. Place pie plate on a wire rack and let set for 10 minutes. Cut into 6 servings.

HINT: Thaw peas by placing in a colander and rinsing under hot water for one minute.

Each serving equals:

HE: 1⅔ Protein (½ limited) • ⅔ Bread •
½ Skim Milk • ¼ Vegetable

---

198 Calories • 6 gm Fat • 17 gm Protein •
19 gm Carbohydrate • 618 mg Sodium •
280 mg Calcium • 1 gm Fiber

DIABETIC: 2 Meat • 1 Starch • ½ Skim Milk

Irish Cabbage Rolls

Perhaps it's my mostly Irish ancestry that inspires me to create healthy versions of Irish classics like this dish that combines a passion for corned beef and cabbage with a new take on stuffed cabbage rolls! I can feel me brogue getting stronger with every bite!

● Serves 4 (2 each)

½ cup chopped onion

1 cup shredded carrots

2 (2.5-ounce) packages Carl Buddig 90% lean corned beef, shredded

8 large cabbage leaves

2 cups hot water☆

1 cup (2¼ ounces) instant potato flakes

⅓ cup Carnation Nonfat Dry Milk Powder

1 (10¾-ounce) can Healthy Request Cream of Mushroom Soup

1 cup skim milk

¼ teaspoon black pepper

1 teaspoon dried parsley flakes

Preheat oven to 350 degrees. Spray a 9-by-13-inch baking dish with butter-flavored cooking spray. In a large skillet sprayed with butter-flavored cooking spray, sauté onion, carrots, and corned beef for 10 minutes or until vegetables are just tender. Meanwhile, place cabbage leaves in a large saucepan. Pour ½ cup hot water over top. Cover and steam about 5 minutes or until just tender. Drain cabbage leaves. Cool. Meanwhile, stir remaining 1½ cups hot water into corned beef mixture. Bring mixture to a boil. Remove from heat. Add dry potato flakes and dry milk powder. Mix well with a fork until fluffy. Place about ¼ cup potato mixture in center of each cabbage leaf. Roll up, folding sides. Place filled leaves in prepared baking dish. In a small bowl, combine mushroom soup, skim milk, black pepper, and parsley flakes. Evenly spoon soup mixture over cabbage rolls. Bake for 25 to 30 minutes. Divide into 4 servings.

Each serving equals:

HE: 1¼ Protein • 1 Vegetable • ¾ Bread •
½ Skim Milk • ½ Slider • 1 Optional Calorie

192 Calories • 4 gm Fat • 13 gm Protein •
26 gm Carbohydrate • 859 mg Sodium •
211 mg Calcium • 2 gm Fiber

DIABETIC: 1 Meat • 1 Starch • 1 Vegetable •
½ Skim Milk

Deluxe Creamed Chipped Beef Casserole

Okay, perhaps not every man will turn away from a football game on TV when you hold a plate of creamed chipped beef under his nose—but most will, including my truck-drivin' man! Here's a terrific casserole version of this man-pleasing classic, definitely worth cheering about! ☻ Serves 6

1 (4.5-ounce) package Hormel 95% lean chipped beef,
    rinsed and shredded
¼ cup finely minced onion
2 cups skim milk
3 tablespoons all-purpose flour
1 (8-ounce) package Philadelphia fat-free cream cheese
½ cup (one 2.5-ounce jar) sliced mushrooms, drained
¾ cup (3 ounces) shredded Kraft reduced-fat Cheddar cheese
2 teaspoons dried parsley flakes
2½ cups hot cooked noodles, rinsed and drained

Preheat oven to 350 degrees. Spray an 8-by-8-inch baking dish with butter-flavored cooking spray. In a large skillet sprayed with butter-flavored cooking spray, sauté chipped beef and onion. In a covered jar, combine skim milk and flour. Shake well to blend. Add milk mixture to beef mixture. Mix well to combine. Continue cooking for 5 minutes or until mixture thickens, stirring often. Add cream cheese, mushrooms, Cheddar cheese, and parsley flakes. Mix well to combine. Stir in noodles. Pour mixture into prepared baking dish. Bake for 25 to 30 minutes. Place baking dish on a wire rack and let set 5 minutes. Divide into 6 servings.

HINT: 2¼ cups uncooked noodles usually cooks to about 2½ cups.

Each serving equals:

HE: 2 Protein • 1 Bread • ⅓ Skim Milk •
¼ Vegetable • 5 Optional Calories

220 Calories • 4 gm Fat • 19 gm Protein •
27 gm Carbohydrate • 603 mg Sodium •
222 mg Calcium • 2 gm Fiber

DIABETIC: 2 Meat • 1½ Starch

Halftime Creamed Chipped Beef

This recipe is so simple to stir up, your husband or sons can actually do it themselves during halftime! It's amazingly creamy and calcium-rich, but contains so little fat per serving, it should top the Hit Parade on every healthy man's list of favorites.

❂ Serves 4 (full ¾ cup)

> **1 recipe JO's Cream Sauce**
> 1 (8-ounce) package Philadelphia fat-free cream cheese
> 1 (4.5-ounce) package Hormel 95% lean chipped beef, rinsed and shredded

In a medium saucepan, combine warm **JO's Cream Sauce** and cream cheese. Mix well to combine. Stir in chipped beef. Continue cooking for 6 to 8 minutes or until cream cheese melts and mixture is heated through, stirring often.

HINT: Good served on toast, potatoes, English muffins, or pasta.

Each serving equals:

HE: 2 Protein • ¾ Skim Milk • ¼ Bread

169 Calories • 1 gm Fat • 23 gm Protein •
17 gm Carbohydrate • 840 mg Sodium •
210 mg Calcium • 0 gm Fiber

DIABETIC: 2 Meat • 1 Skim Milk

Desserts

Strawberry Romanoff Cheesecake Parfait

If you've never tasted the renowned dessert called Strawberries Romanoff, let me just say that this rich concoction of berries, cheesecake, and creamy sauce was for years considered the "crème de la crème" of classy desserts! Now you can enjoy those irresistible flavors at home—and feel good about it, too! ❤ Serves 4

1 (8-ounce) package Philadelphia fat-free cream cheese
1 (4-serving) package JELL-O sugar-free instant vanilla
    pudding mix
⅔ cup Carnation Nonfat Dry Milk Powder
1 cup unsweetened orange juice
¼ cup Cool Whip Free
1 cup sliced fresh strawberries☆

In a large bowl, stir cream cheese with a spoon until soft. Add dry pudding mix, dry milk powder, and orange juice. Mix well using a wire whisk. Blend in Cool Whip Free. Spoon 3 tablespoons of the mixture into 4 parfait dishes. Reserve 4 pieces of strawberry for garnish. Evenly divide remaining strawberries between the 4 parfait dishes. Spoon remaining pudding mixture evenly over the strawberries. Garnish each with a reserved piece of strawberry. Refrigerate for at least 1 hour.

Each serving equals:

HE: 1 Protein • ¾ Fruit • ½ Skim Milk •
¼ Slider • 13 Optional Calories

144 Calories • 0 gm Fat • 12 gm Protein •
24 gm Carbohydrate • 735 mg Sodium •
148 mg Calcium • 1 gm Fiber

DIABETIC: 1 Meat • 1 Fruit • ½ Skim Milk •
½ Starch/Carbohydrate

Pears au Chocolate

I love reading menus, especially descriptions of tasty desserts created by chefs. When I first heard of a dish that "married" pears to chocolate, I knew I must invent a Healthy Exchanges version! The touch of mint gives this romantic treat a little extra glamour—don't you agree?   ☻   Serves 4

> 2 cups (one 16-ounce can) pear halves, packed in fruit juice,
>   drained, and ½ cup liquid reserved
> 1 (4-serving) package JELL-O sugar-free chocolate
>   cook-and-serve pudding mix
> ⅔ cup Carnation Nonfat Dry Milk Powder
> 1 cup water
> ¼ teaspoon peppermint extract
> ¼ cup Cool Whip Free
> 2 maraschino cherries, halved

Cut pear halves in half and evenly divide pieces among 4 dessert dishes. In a medium saucepan, combine dry pudding mix and dry milk powder. Add reserved pear liquid and water. Mix well to combine. Cook over medium heat for 6 to 8 minutes or until mixture thickens and starts to boil, stirring constantly. Remove from heat. Stir in peppermint extract. Evenly spoon hot mixture over pears. Cover and refrigerate for at least 30 minutes. When serving, top each with 1 tablespoon Cool Whip Free and ½ maraschino cherry.

Each serving equals:

HE: 1 Fruit • ½ Skim Milk • ¼ Slider •
18 Optional Calories

172 Calories • 0 gm Fat • 5 gm Protein •
38 gm Carbohydrate • 180 mg Sodium •
144 mg Calcium • 2 gm Fiber

DIABETIC: 1 Fruit • 1 Starch • ½ Skim Milk

Fruit and "Custard" Pudding

Here's a light and pretty dessert you can do for a luncheon or afternoon tea party. The varied fruit flavors and textures will please your taste buds, and the oh-so-cool and creamy "custard" will delight all your senses.   ❂   Serves 8

> 1 (4-serving) package JELL-O sugar-free vanilla
>   cook-and-serve pudding mix
> ⅔ cup Carnation Nonfat Dry Milk Powder
> 1 cup water
> ½ cup unsweetened orange juice
> 1 (8-ounce) package Philadelphia fat-free cream cheese
> 1 cup (one medium) diced banana
> 1 cup (one 11-ounce can) mandarin oranges,
>   rinsed and drained
> 1½ cups (9 ounces) seedless white grapes

In a medium saucepan, combine dry pudding mix, dry milk powder, and water. Stir in orange juice. Cook over medium heat for 5 minutes or until mixture thickens and starts to boil, stirring constantly. Remove from heat. Add cream cheese. Mix well using a wire whisk. Fold in banana, mandarin oranges, and grapes. Evenly spoon mixture into 8 dessert dishes. Refrigerate for at least 30 minutes.

HINT: To prevent bananas from turning brown, mix with 1 teaspoon lemon juice or sprinkle with Fruit Fresh.

Each serving equals:

> HE: 1 Fruit • ½ Protein • ¼ Skim Milk •
> 10 Optional Calories
>
> ---
>
> 112 Calories • 0 gm Fat • 7 gm Protein •
> 21 gm Carbohydrate • 261 mg Sodium •
> 78 mg Calcium • 1 gm Fiber
>
> ---
>
> DIABETIC: 1 Fruit • ½ Meat • ½ Starch/Carbohydrate

Tropical Treasure Pudding ❄

Sure, it might seem simpler to stir up a package of pudding and plop it into dessert dishes to serve your family. But with just a bit more effort, a dash of spice, and the magic of dry milk powder—you've got a dessert worth following a map to discover! My grandson Joshie likes to smack his lips over this one!   ☺   Serves 4

> 1 (4-serving) package JELL-O sugar-free instant banana
>   pudding mix
> ⅔ cup Carnation Nonfat Dry Milk Powder
> 1 cup water
> 1 cup (one 8-ounce can) crushed pineapple, packed in
>   fruit juice, undrained
> 1 cup (one medium) diced banana
> ½ cup Cool Whip Free
> ½ teaspoon apple pie spice

In a large bowl, combine dry pudding mix, dry milk powder, and water. Mix well using a wire whisk. Stir in undrained pineapple and banana. Add Cool Whip Free and apple pie spice. Mix gently to combine. Evenly spoon mixture into 4 dessert dishes. Refrigerate for at least 30 minutes.

HINT: To prevent banana from turning brown, mix with 1 teaspoon lemon juice or sprinkle with Fruit Fresh.

Each serving equals:

HE: 1 Fruit • ½ Skim Milk • ½ Slider •
1 Optional Calorie

156 Calories • 0 gm Fat • 5 gm Protein •
34 gm Carbohydrate • 408 mg Sodium •
149 mg Calcium • 1 gm Fiber

DIABETIC: 1 Fruit • ½ Starch • ½ Skim Milk
or 1 Fruit • 1 Starch/Carbohydrate

Raspberry Chocolate Jewels

The fresh raspberry season isn't very long, so you need to be ready to take advantage of every delicious moment! This pretty dessert looks as if you really fussed, but it takes only a few minutes to make it sparkle. ◑ Serves 4

1 (4-serving) package JELL-O sugar-free instant chocolate pudding mix
⅓ cup Carnation Nonfat Dry Milk Powder
1½ cups Yoplait plain fat-free yogurt
½ cup water
½ teaspoon almond extract
6 (2½-inch) chocolate graham crackers, made into crumbs
¾ cup fresh red raspberries
¼ cup Cool Whip Free

In a medium bowl, combine dry pudding mix and dry milk powder. Add yogurt and water. Mix well using a wire whisk. Blend in almond extract. Spoon about ¼ cup pudding mixture into 4 dessert or parfait dishes. Spoon 1 tablespoon graham cracker crumbs over top of each. Reserve 4 raspberries. Evenly divide remaining raspberries among the four dishes. Spoon remaining chocolate pudding (about ¼ cup each) evenly over raspberries. Top each with 1 tablespoon Cool Whip Free. Sprinkle about 1½ teaspoons crumbs over top of each and garnish with 1 reserved raspberry. Refrigerate for at least 1 hour.

Each serving equals:

HE: ¾ Skim Milk • ½ Bread • ¼ Fruit •
¼ Slider • 13 Optional Calories

137 Calories • 1 gm Fat • 8 gm Protein •
24 gm Carbohydrate • 459 mg Sodium •
244 mg Calcium • 1 gm Fiber

DIABETIC: 1 Skim Milk • 1 Starch/Carbohydrate

Rocky Road Rewards

I confess it: I get some of my favorite dessert ideas from delectable ice cream flavors and their mix-in treats! Whoever created the first rocky road deserves a prize, but perhaps they'd settle for this healthy and tasty version that brims with treats like walnuts and marshmallows. ☻ Serves 4

> 1 (4-serving) package JELL-O sugar-free instant chocolate
>   pudding mix
> 1 cup skim milk
> ¾ cup Yoplait plain fat-free yogurt
> 1 teaspoon vanilla extract
> 2 tablespoons (½ ounce) chopped walnuts
> 1 tablespoon (¼ ounce) mini chocolate chips
> ¼ cup (½ ounce) miniature marshmallows
> ¼ cup Cool Whip Free

In a medium bowl, combine dry pudding mix, skim milk, and yogurt. Mix well using a wire whisk. Blend in vanilla extract, walnuts, chocolate chips, and marshmallows. Fold in Cool Whip Free. Evenly spoon mixture into 4 dessert dishes. Refrigerate for at least 30 minutes.

Each serving equals:

HE: ½ Skim Milk • ¼ Fat • ¾ Slider •
7 Optional Calories

124 Calories • 3 gm Fat • 6 gm Protein •
18 gm Carbohydrate • 395 mg Sodium •
164 mg Calcium • 0 gm Fiber

DIABETIC: 1 Starch/Carbohydrate • ½ Skim Milk •
½ Fat

Mocha Almond Mousse

It's fudgy, it's creamy, it's nutty—and now it's yours to serve anytime you like! If you're a fan of unusual coffee blends, you may want to experiment using coffees from around the world to flavor this luscious too-good-to-be-just-pudding!   ☻   Serves 4

> 1 (4-serving) package JELL-O sugar-free instant
>   chocolate fudge pudding mix
> **1 recipe JO's Sour Cream**
> 1 cup cold coffee
> 1 teaspoon almond extract
> ¼ cup Cool Whip Free
> 2 tablespoons (½ ounce) chopped almonds

In a medium bowl, combine dry pudding mix, **JO's Sour Cream**, and cold coffee. Mix well using a wire whisk. Blend in almond extract. Evenly spoon mixture into 4 dessert dishes. Spoon 1 tablespoon Cool Whip Free and sprinkle 1½ teaspoons almonds over top of each. Refrigerate for at least 30 minutes.

Each serving equals:

HE: ½ Skim Milk • ¼ Fat • ½ Slider •
10 Optional Calories

102 Calories • 2 gm Fat • 6 gm Protein •
15 gm Carbohydrate • 394 mg Sodium •
166 mg Calcium • 1 gm Fiber

DIABETIC: 1 Starch/Carbohydrate • ½ Fat

Fruit Cocktail Rice Pudding

Most men and kids I know love rice pudding—and so do I! That's why I like devising new and exciting ways to serve this old-fashioned favorite. This recipe would be just about perfect with the fruit cocktail and pistachio pudding blended together, but the coconut extract makes it magical. ☻ Serves 6

1 (4-serving) package JELL-O sugar-free instant pistachio
    pudding mix
⅔ cup Carnation Nonfat Dry Milk Powder
2 cups (one 16-ounce can) fruit cocktail, packed in fruit juice,
    drained and ⅓ cup liquid reserved
1 cup water
1 teaspoon coconut extract
¼ cup Cool Whip Free
1½ cups cold cooked rice

In a large bowl, combine dry pudding mix and dry milk powder. Add reserved fruit cocktail liquid and water. Mix well using a wire whisk. Stir in coconut extract and Cool Whip Free. Fold in rice and fruit cocktail. Evenly spoon mixture into 6 dessert dishes. Refrigerate for at least 30 minutes.

HINT: 1 cup uncooked rice usually cooks to about 1½ cups.

Each serving equals:

HE: ⅔ Fruit • ½ Bread • ⅓ Skim Milk •
¼ Slider • 2 Optional Calories

---

128 Calories • 0 gm Fat • 4 gm Protein •
28 gm Carbohydrate • 261 mg Sodium •
102 mg Calcium • 1 gm Fiber

DIABETIC: 1 Starch/Carbohydrate • ½ Fruit

Banana Maple Rice Pudding

Bananas topped with maple syrup sound like a wonderful idea for topping pancakes or French toast. But why not take that tasty treat and stir it into a rice pudding dessert? The crunchy pecans will convince your mouth you're dining New Orleans–style on pralines! John, the family dessert specialist, calls this one of his favorites.

○ Serves 6 (full ¾ cup)

> 1 (4-serving) package JELL-O sugar-free instant banana
>   pudding mix
> **2 recipes JO's Evaporated Skim Milk**
> ½ cup Cary's Sugar Free Maple Syrup
> ¾ cup Cool Whip Free
> 2 cups (2 medium) diced bananas
> 3 tablespoons (¾ ounce) chopped pecans
> 2 cups cold cooked rice

In a large bowl, combine dry pudding mix and **JO's Evaporated Skim Milk**. Mix well using a wire whisk. Stir in maple syrup. Blend in Cool Whip Free. Add bananas, pecans, and rice. Mix gently to combine. Evenly spoon mixture into 6 dessert dishes. Refrigerate for at least 15 minutes.

HINTS: 1. To prevent bananas from turning brown, mix with 1 teaspoon lemon juice or sprinkle with Fruit Fresh.
2. 1⅓ cups uncooked rice usually cooks to about 2 cups.

Each serving equals:

HE: ⅔ Fruit • ⅔ Bread • ½ Fat • ⅓ Skim Milk • ¼ Slider • 10 Optional Calories

199 Calories • 3 gm Fat • 4 gm Protein •
38 gm Carbohydrate • 296 mg Sodium •
100 mg Calcium • 1 gm Fiber

DIABETIC: 1 Fruit • 1 Starch/Carbohydrate • ½ Fat

Peach Bread Pudding with Blueberry Sauce

❄

I enjoy bread pudding every chance I get, but because the traditional version of this dessert is often made with cream, I most often enjoy it the Healthy Exchanges Way. This one combines two of my favorite fruits—peaches and blueberries—and because it uses them canned and frozen, you can savor this all year long.

♥ Serves 6

> 2 cups (one 16-ounce can) sliced peaches,
>   packed in fruit juice, drained, coarsely chopped,
>   and ⅓ cup liquid reserved
> 8 slices reduced-calorie white bread, torn into pieces
> 1 (4-serving) package JELL-O sugar-free vanilla
>   cook-and-serve pudding mix
> 2 cups skim milk
> ½ teaspoon ground nutmeg
> 1 teaspoon vanilla extract
> ⅔ cup water
> 1½ cups frozen unsweetened blueberries, thawed, and
>   ¼ cup liquid reserved
> 1 tablespoon cornstarch
> 2 tablespoons Sugar Twin or Sprinkle Sweet

Preheat oven to 350 degrees. Spray an 8-by-8-inch baking dish with butter-flavored cooking spray. In a large bowl, combine drained peaches and bread. In a small bowl, combine dry pudding mix, skim milk, nutmeg, and vanilla extract. Add pudding mixture to bread mixture. Mix gently to combine. Pour mixture into prepared baking dish. Bake for 30 to 35 minutes. Meanwhile, in a medium saucepan, combine reserved fruit liquids, water, cornstarch, and Sugar Twin. Cook over medium heat for 5 minutes or until mixture thickens, stirring constantly. Gently stir in blueberries. Lower heat and simmer just to keep sauce warm. When serving, cut bread pudding into 6 servings and spoon about ¼ cup warm sauce over top of each.

Each serving equals:

HE: 1 Fruit • ⅔ Bread • ⅓ Skim Milk • ¼ Slider

---

164 Calories • 0 gm Fat • 7 gm Protein • 34 gm Carbohydrate • 278 mg Sodium • 131 mg Calcium • 5 gm Fiber

---

DIABETIC: 1 Fruit • 1 Starch/Carbohydrate

Fruit Noodle Pudding

A savory dish that came to America's shores from Eastern Europe, noodle pudding is a dish of many contrasts and flavors. This recipe blends in more fruit than most to produce a truly scrumptious and satisfying dessert. It makes a lovely addition to the table for a weekend brunch buffet. ☻ Serves 8

> 1 cup (one 8-ounce can) crushed pineapple, packed in
>   fruit juice, drained, and ⅓ cup liquid reserved
> 1 cup (one 8-ounce can) sliced peaches, packed in
>   fruit juice, drained, and ⅓ cup liquid reserved
> 1 cup water
> 1 (4-serving) package JELL-O sugar-free vanilla
>   cook-and-serve pudding mix
> ⅔ cup Carnation Nonfat Dry Milk Powder
> 1 teaspoon vanilla extract
> 2 tablespoons Brown Sugar Twin
> ½ teaspoon apple pie spice
> ¼ cup raisins
> ⅓ cup (1½ ounces) chopped walnuts
> 2 cups cold cooked noodles, rinsed and drained

Preheat oven to 350 degrees. Spray an 8-by-8-inch baking dish with butter-flavored cooking spray. In a large saucepan, combine reserved fruit liquids, water, dry pudding mix, and dry milk powder. Mix well using a wire whisk. Cook over medium heat for 6 to 8 minutes or until mixture thickens and starts to boil, stirring constantly. Remove from heat. Stir in vanilla extract, Brown Sugar Twin, and apple pie spice. Add pineapple, peaches, raisins, walnuts, and noodles. Mix well to combine. Spread mixture into prepared baking dish. Bake for 30 minutes. Place baking dish on a wire rack and let set 5 minutes. Divide into 8 servings. Good warm or cold.

HINT: 1¾ cups uncooked noodles usually cooks to about
    2 cups.

Each serving equals:

HE: 1 Fruit • ⅔ Bread • ½ Fat • ¼ Protein •
15 Optional Calories

---

168 Calories • 4 gm Fat • 5 gm Protein •
28 gm Carbohydrate • 94 mg Sodium •
87 mg Calcium • 2 gm Fiber

DIABETIC: 1 Fruit • 1 Starch/Carbohydrate • ½ Fat

Chocolate Strawberry Shortcake

For all those who can't decide which dessert flavor is their favorite, chocolate or strawberry, here's a dazzling creation that celebrates both! If you've never made a cake with mayonnaise in it, you'll be pleased how moist the finished cake tastes. ☯ Serves 4

4 recipes JO's Chocolate Milk Beverage Mix
¾ *cup all-purpose flour*
1 tablespoon baking powder
¼ *cup Kraft fat-free mayonnaise*
1½ cups water
4 cups sliced fresh strawberries☆
Sugar substitute to equal 2 tablespoons sugar
¼ *cup Cool Whip Lite*
4 teaspoons Hershey's Lite Chocolate Syrup

Preheat oven to 350 degrees. Spray an 8-by-8-inch baking dish with butter-flavored cooking spray. In a medium bowl, combine dry **JO's Chocolate Milk Beverage Mix**, flour, and baking powder. Add mayonnaise and water. Mix well to combine. Pour mixture into a prepared baking dish. Bake for 20 to 25 minutes or until a toothpick inserted in center comes out clean. Place baking dish on a wire rack and allow to cool. Meanwhile, in a medium bowl, crush 1 cup strawberries with a potato masher or fork. Add sugar substitute. Mix well to combine. Stir in remaining strawberries. Cover and refrigerate. When ready to serve, cut shortcake into 4 pieces. For each serving, place 1 piece of shortcake on a dessert plate, spoon about ¾ cup strawberry sauce over top and garnish with 1 tablespoon Cool Whip Lite and 1 teaspoon chocolate syrup.

Each serving equals:

HE: 1 Skim Milk • 1 Bread • 1 Fruit • ½ Slider •
5 Optional Calories

266 Calories • 2 gm Fat • 12 gm Protein •
50 gm Carbohydrate • 628 mg Sodium •
508 mg Calcium • 4 gm Fiber

DIABETIC: 1½ Starch/Carbohydrate • 1 Skim Milk •
1 Fruit

Yankee Doodle Dandy's Shortcakes

Three cheers for the red, white, and blue—especially when you can be patriotic by choosing this tri-color dessert to provide a summer picnic's fireworks finale! What a beautiful way to show your support for the good old USA!   ☻   Serves 4

¾ cup Bisquick Reduced Fat Baking Mix
⅓ cup Carnation Nonfat Dry Milk Powder
½ cup Sugar Twin or Sprinkle Sweet☆
2 tablespoons Kraft fat-free mayonnaise
⅓ cup water
1 teaspoon vanilla extract
2 cups fresh sliced strawberries
1 cup (one medium) diced banana
¾ cup fresh blueberries
¼ cup Cool Whip Lite

Preheat oven to 415 degrees. Spray a cookie sheet with butter-flavored cooking spray. In a medium bowl, combine baking mix, dry milk powder, and ¼ cup Sugar Twin. Add mayonnaise, water, and vanilla extract. Mix well to combine. Drop by spoonfuls onto prepared cookie sheet, to form 4 shortcakes. Bake for 8 to 12 minutes or until golden brown. Place cookie sheet on a wire rack and allow to cool. Meanwhile, in a medium bowl, mash 1 cup sliced strawberries with a potato masher or fork. Stir in remaining ¼ cup Sugar Twin. Add remaining strawberries, banana, and blueberries. Mix gently to combine. Cover and refrigerate until ready to serve. For each serving, place 1 piece of shortcake on a dessert plate, spoon about ¾ cup fruit mixture over top, and garnish with 1 tablespoon Cool Whip Lite.

Each serving equals:

HE: 1¼ Fruit • 1 Bread • ¼ Skim Milk •
¼ Slider • 5 Optional Calories

194 Calories • 2 gm Fat • 5 gm Protein •
39 gm Carbohydrate • 360 mg Sodium •
101 mg Calcium • 3 gm Fiber

DIABETIC: 1½ Starch/Carbohydrate • 1 Fruit • ½ Fat

September Song Shortcakes

How does that old song lyric go about saying good-bye to the summer? It's never easy to face the return of fall and back-to-school time, but this sweet harvest dessert may soften the blow. There's something wonderful about the combination of apples and pecans, a hint of Indian summer in the air—don't you agree?

Serves 8

1½ cups Bisquick Reduced Fat Baking Mix
⅔ cup Carnation Nonfat Dry Milk Powder
1 teaspoon apple pie spice☆
⅓ cup Sugar Twin or Sprinkle Sweet
¼ cup Kraft fat-free mayonnaise
1⅔ cups water☆
1 teaspoon vanilla extract
¼ cup raisins
1 cup unsweetened apple juice
1 (4-serving) package JELL-O sugar-free vanilla
   cook-and-serve pudding mix
2 cups (4 small) cored and chopped cooking apples
2 tablespoons (½ ounce) chopped pecans

Preheat oven to 415 degrees. Spray 8 wells of a 12-hole muffin pan with butter-flavored cooking spray or line with paper liners. In a large bowl, combine baking mix, dry milk powder, ½ teaspoon apple pie spice, and Sugar Twin. Add mayonnaise, ⅔ cup water, and vanilla extract. Mix well to combine. Stir in raisins. Evenly spoon batter into prepared muffin pan. Bake for 10 to 12 minutes or until tops are lightly browned. Place muffin pan on a wire rack. Meanwhile in a medium saucepan, combine remaining 1 cup water, apple juice, and dry pudding mix. Stir in remaining ½ teaspoon apple pie spice and apples. Cook over medium heat for 10 minutes or until apples soften and mixture starts to thicken, stirring often. Remove from heat. For each serving, place a shortcake on a dessert

dish, spoon a scant ⅓ cup apple mixture over top and garnish with
¾ teaspoon pecans.

HINTS: 1. Fill unused muffin wells with water. It protects the
            muffin tin and ensures even baking.
        2. Rewarm leftover apple mixture in microwave.
        3. Good warm or cold.
        4. Good served with ½ cup sugar- and fat-free vanilla
            ice cream.

Each serving equals:

HE: 1 Bread • 1 Fruit • ¼ Skim Milk • ¼ Fat •
19 Optional Calories

---

179 Calories • 3 gm Fat • 4 gm Protein •
34 gm Carbohydrate • 416 mg Sodium •
94 mg Calcium • 1 gm Fiber

DIABETIC: 1½ Starch/Carbohydrate • 1 Fruit • ½ Fat

Chocolate-Maple Brandy Balls ❄

As the holidays approach, it's important not to feel deprived of seasonal goodies. Now, with Healthy Exchanges treats like this one, you never have to feel that way again! Remember, brandy extract contains no real liquor, so even your children can enjoy these tasty bite-size balls. ☻ Serves 6 (2 each)

1 (4-serving) package JELL-O sugar-free instant chocolate
    pudding mix
⅔ cup Carnation Nonfat Dry Milk Powder
6 tablespoons purchased graham cracker crumbs or
    6 (2½-inch) graham crackers, made into fine crumbs
6 tablespoons Cary's Sugar Free Maple Syrup
1 teaspoon brandy extract

In a medium bowl, combine dry pudding mix, dry milk powder, and graham cracker crumbs. Add maple syrup and brandy extract. Mix gently to combine. Place waxed paper on a cookie sheet. Roll mixture into 12 balls and arrange on cookie sheet. Cover and refrigerate for at least 30 minutes.

HINTS: 1. A sandwich bag works great for making graham cracker crumbs.
         2. If mixture is too dry, add a few drops of water at a time until mixture is the right consistency to roll into balls.

Each serving equals:

HE: ⅓ Skim Milk • ⅓ Bread • ¼ Slider •
7 Optional Calories

84 Calories • 0 gm Fat • 4 gm Protein •
17 gm Carbohydrate • 340 mg Sodium •
94 mg Calcium • 0 gm Fiber

DIABETIC: 1 Starch/Carbohydrate

Oatmeal Nut Drops

Do you usually participate in a holiday cookie exchange—where you trade cookies with other holiday bakers—but figured you wouldn't be able to this year? These nutty treats don't taste like "diet food," so you can feel comfortable, even proud, to share them with friends and family. You're passing along the gift of good health.

● Serves 6 (4 each)

2 tablespoons Peter Pan reduced-fat peanut butter
¼ cup Cary's Sugar Free Maple Syrup
¼ cup Sugar Twin or Sprinkle Sweet
⅔ cup Carnation Nonfat Dry Milk Powder
1 cup (3 ounces) quick oats
¼ cup (1 ounce) chopped dry-roasted peanuts

In a medium bowl, combine peanut butter, maple syrup, Sugar Twin, and dry milk powder. Stir in oats and peanuts. Place waxed paper on a cookie sheet. Drop mixture by teaspoonfuls onto cookie sheet to form 24 drops. Cover and refrigerate for at least 30 minutes.

Each serving equals:

HE: ⅔ Bread • ⅔ Fat • ½ Protein • ⅓ Skim Milk • 11 Optional Calories

---

166 Calories • 6 gm Fat • 8 gm Protein • 20 gm Carbohydrate • 90 mg Sodium • 104 mg Calcium • 2 gm Fiber

---

DIABETIC: 1 Starch/Carbohydrate • 1 Fat • ½ Meat

Individual Brownie Alaska

If you've tasted Baked Alaska in a restaurant but never dared try that meringue-topped creation at home, here's your chance! This dish features lots of scrumptious flavors even before you get to those creamy peaks. Angie, Tommy's "friend," decided that healthy was just fine as long as it tasted this good!

○ Serves 2

> **2 recipes JO's Chocolate Milk Beverage Mix**
> ½ cup (1½ ounces) quick oats
> ½ teaspoon baking powder
> 2 tablespoons Peter Pan reduced-fat peanut butter
> 1 teaspoon coconut extract☆
> ½ cup water
> 2 egg whites
> Sugar substitute to equal 2 tablespoons sugar
> ⅔ cup sugar and fat-free vanilla ice cream

Preheat oven to 350 degrees. Spray two (12-ounce) custard dishes with butter-flavored cooking spray. In a medium bowl, combine dry **JO's Chocolate Milk Beverage Mix**, oats, and baking powder. Add peanut butter, ½ teaspoon coconut extract, and water. Mix well to combine. Evenly spread mixture into prepared dishes. Bake for 12 to 15 minutes or until a toothpick inserted in center comes out clean. Place baking dishes on a wire rack and allow to cool. Reset oven to 450 degrees. Meanwhile, in a medium bowl, beat egg whites with an electric mixer until soft peaks form. Add sugar substitute and remaining ½ teaspoon coconut extract. Continue beating until stiff peaks form. Place ⅓ cup hard ice cream on top of each slightly cooled brownie. Spread meringue mixture evenly over ice cream, being sure to seal to edge of dish. Bake for 4 to 5 minutes at 450 degrees or until meringue starts to turn brown. Serve at once.

Each serving equals:

HE: 1 Skim Milk • 1 Bread • 1 Protein • 1 Fat •
½ Slider • 9 Optional Calories

---

339 Calories • 7 gm Fat • 22 gm Protein •
47 gm Carbohydrate • 409 mg Sodium •
440 mg Calcium • 4 gm Fiber

DIABETIC: 1½ Starch/Carbohydrate • 1 Skim Milk •
1 Fat • ½ Meat

Banana Graham Cracker Dessert

Don't have a piecrust handy but want a delectable dessert that "eats" like pie? Here's how! You simply arrange graham cracker squares in your cake pan, then spread your topping over it. Just wait until you taste the terrific result!   ☺   Serves 8

> 12 (2½-inch) graham cracker squares☆
>
> **1 recipe JO's Deliteful Whipped Topping**
>
> 1 teaspoon rum extract
>
> 3 cups (3 medium) sliced bananas
>
> ¼ cup (1 ounce) chopped pecans
>
> 1 tablespoon Brown Sugar Twin

Evenly arrange 9 graham cracker squares in the bottom of a 9-by-9-inch cake pan. In a medium bowl, combine **JO's Deliteful Whipped Topping** and rum extract. Evenly spread ⅓ of topping mixture over graham crackers. Layer bananas evenly over top. Evenly spread remaining topping mixture over bananas. Crush remaining 3 graham crackers into fine crumbs. In a small bowl, combine graham cracker crumbs, pecans, and Brown Sugar Twin. Sprinkle crumb mixture evenly over top. Cover and refrigerate for at least 2 hours. Cut into 8 servings.

HINT: To prevent bananas from turning brown, mix with 1 teaspoon lemon juice or sprinkle with Fruit Fresh.

Each serving equals:

HE: ¾ Fruit • ½ Bread • ½ Fat • ¼ Skim Milk • ¼ Slider • 4 Optional Calories

---

139 Calories • 3 gm Fat • 3 gm Protein • 25 gm Carbohydrate • 71 mg Sodium • 82 mg Calcium • 1 gm Fiber

---

DIABETIC: 1 Fruit • ½ Starch/Carbohydrate • ½ Fat

Lemon Crunch Bars

These flavorful cookie bars are sweet and crunchy, tangy and tart all at once. And better still, you get to nibble two of them for one satisfying serving! ♥ Serves 12 (2 each)

1 (8-ounce) can Pillsbury Reduced Fat Crescent Rolls
1 (4-serving) package JELL-O sugar-free vanilla
    cook-and-serve pudding mix
1 (4-serving) package JELL-O sugar-free lemon gelatin
1 cup Carnation Nonfat Dry Milk Powder
1½ cups water
1 teaspoon coconut extract
1 (8-ounce) package Philadelphia fat-free cream cheese
1½ cups raisins
¼ cup (1 ounce) chopped pecans
¼ cup flaked coconut

Preheat oven to 425 degrees. Spray a 9-by-13-inch rimmed cookie sheet with butter-flavored cooking spray. Pat rolls in pan, being sure to seal perforations. Bake for 6 to 8 minutes or until light golden brown. Place cookie sheet on a wire rack and allow to cool. Meanwhile, in a medium saucepan, combine dry pudding mix, dry gelatin, dry milk powder, and water. Cook over medium heat for 6 to 8 minutes or until mixture thickens and starts to boil, stirring constantly. Remove from heat. Add coconut extract and cream cheese. Mix well using a wire whisk. Fold in raisins, pecans, and coconut. Spread mixture evenly over cooled crust. Refrigerate for at least 2 hours. Cut into 24 bars.

HINT: Do not use inexpensive rolls as they don't cover the pan properly.

Each serving equals:

HE: 1 Fruit • ⅔ Bread • ⅓ Protein • ⅓ Fat •
¼ Skim Milk • 15 Optional Calories

189 Calories • 5 gm Fat • 7 gm Protein •
29 gm Carbohydrate • 355 mg Sodium •
80 mg Calcium • 1 gm Fiber

DIABETIC: 1 Fruit • 1 Starch/Carbohydrate • 1 Fat

May Day Lemon Strawberry Dessert

Once the market boasts the first red, ripe jewels in late spring, I bring out all my best strawberry sweets! This one takes a few minutes to assemble, but when you're done, it's a dazzler! So easy and so elegant. ☺ Serves 8

2 cups fresh strawberries

3 tablespoons Sugar Twin or Sprinkle Sweet☆

12 (2½-inch) graham cracker squares☆

2 (4-serving) packages JELL-O sugar-free lemon gelatin

1 (4-serving) package JELL-O sugar-free instant vanilla
   pudding mix

⅔ cup Carnation Nonfat Dry Milk Powder

1½ cups Yoplait plain fat-free yogurt

1 cup Cool Whip Free

1 teaspoon almond extract

¼ cup (1 ounce) sliced almonds

Reserve 4 whole strawberries. Slice remaining strawberries. In a medium bowl, combine sliced strawberries and 2 tablespoons Sugar Twin. Set aside. Evenly arrange 9 graham cracker squares in the bottom of a 9-by-9-inch cake pan. In a medium bowl, combine dry gelatin, dry pudding mix, and dry milk powder. Add yogurt. Mix well using a wire whisk. Blend in Cool Whip Free and almond extract. Spread half of pudding mixture evenly over graham crackers. Top with sliced strawberries. Spread remaining pudding mixture over strawberries. Refrigerate. Meanwhile, crush remaining 3 graham crackers into fine crumbs. In a small skillet sprayed with butter-flavored cooking spray, combine graham cracker crumbs, remaining 1 tablespoon Sugar Twin, and almonds. Cook over medium heat for 2 to 3 minutes or until mixture is lightly toasted, stirring constantly. Cool completely. Evenly sprinkle mixture over top of dessert. Cut reserved strawberries in half and garnish top with strawberry halves. Cover and refrigerate for at least 2 hours. Cut into 8 servings.

Each serving equals:

HE: ½ Bread • ½ Skim Milk • ¼ Fat • ¼ Fruit •
½ Slider • 7 Optional Calories

---

130 Calories • 2 gm Fat • 7 gm Protein •
21 gm Carbohydrate • 322 mg Sodium •
168 mg Calcium • 1 gm Fiber

DIABETIC: 1 Starch/Carbohydrate • ½ Skim Milk • ½ Fat
or 1½ Starch/Carbohydrate • ½ Fat

Cupid's Chocolate Cherry Dessert Bars

Want to tell someone how much you love 'em on Valentine's Day, or simply to say thank you for being so nice? Here's a delightful way to show your affection! ☺ Serves 16

> 1 (8-ounce) can Pillsbury Reduced Fat Crescent Rolls
>
> 2 (4-serving) packages JELL-O sugar-free chocolate cook-and-serve pudding mix
>
> 1 (4-serving) package JELL-O sugar-free cherry gelatin
>
> 1 cup Carnation Nonfat Dry Milk Powder☆
>
> 2 cups (one 16-ounce can) tart red cherries, packed in water, drained, and ¼ cup liquid reserved
>
> 2½ cups water
>
> 1 (8-ounce) package Philadelphia fat-free cream cheese
>
> ½ cup Sugar Twin or Sprinkle Sweet☆
>
> 1½ teaspoons almond extract☆
>
> 1¾ cups Yoplait plain fat-free yogurt
>
> 1½ cups Cool Whip Free
>
> ¼ cup (1 ounce) chopped almonds

Preheat oven to 425 degrees. Spray a rimmed 9-by-13-inch cookie sheet with butter-flavored cooking spray. Pat rolls into pan, being sure to seal perforations. Bake for 6 to 8 minutes or until light golden brown. Place cookie sheet on a wire rack and allow to cool. Meanwhile, in a medium saucepan, combine dry pudding mixes, dry gelatin, and ⅔ cup dry milk powder. Add reserved cherry liquid and water. Mix well to combine. Stir in cherries. Cook over medium heat for 6 to 8 minutes or until mixture thickens and starts to boil, stirring constantly and being sure not to crush cherries. Remove from heat. Place pan on a wire rack and allow to cool for 15 minutes, stirring occasionally. Meanwhile, in a medium bowl, stir cream cheese with a spoon until soft. Add 2 tablespoons Sugar Twin and ½ teaspoon almond extract. Mix well to combine. Spread cream cheese mixture evenly over cooled crust. Spread cooled cherry mixture evenly over cream cheese mixture. Refrigerate for at least 30 minutes. In a medium bowl, combine yogurt and remaining ⅓ cup dry milk powder. Blend in remaining 6 tablespoons Sugar Twin and remaining 1 teaspoon almond extract. Add Cool Whip Free. Mix gently to combine. Spread topping mixture evenly over cherry mixture. Sprinkle chopped almonds evenly over top. Refrigerate for at least 1 hour. Cut into 16 pieces.

HINT: Do not use inexpensive rolls as they don't cover the pan properly.

Each serving equals:

HE: ½ Bread • ⅓ Protein • ⅓ Skim Milk •
¼ Fruit • ¼ Slider • 9 Optional Calories

---

135 Calories • 3 gm Fat • 7 gm Protein •
20 gm Carbohydrate • 312 mg Sodium •
111 mg Calcium • 1 gm Fiber

DIABETIC: 1 Starch/Carbohydrate • ½ Fat

Death by Chocolate

This recipe name may sound a little outrageous, but then so is this ultra-chocolatey dessert! Take comfort in this, at least—if you could "die" by chocolate, you'd leave with a smile on your face. (Can you believe the original version of this recipe had 52 grams of fat per serving????)　●　Serves 16 (full ½ cup)

> 1½ cups all-purpose flour
> ¼ cup unsweetened cocoa
> ½ teaspoon baking soda
> 1 teaspoon baking powder
> ½ cup + 2 tablespoons Sugar Twin or Sprinkle Sweet☆
> ¾ cup Kraft fat-free mayonnaise
> 2½ cups water☆
> 3½ teaspoons vanilla extract☆
> ⅓ cup Hershey's Lite Chocolate Syrup
> 2 (4-serving) packages JELL-O sugar-free instant
>   chocolate fudge pudding mix
> 1⅓ cups Carnation Nonfat Dry Milk Powder☆
> 3 cups Yoplait plain fat-free yogurt☆
> 2 cups Cool Whip Free
> ¼ cup (1 ounce) Heath Bits Toffee Chips

Preheat oven to 350 degrees. Spray an 11-by-7-inch baking pan with butter-flavored cooking spray. In a large bowl, combine flour, cocoa, baking soda, baking powder, and ½ cup Sugar Twin. In a small bowl, combine mayonnaise, 1 cup water and 1½ teaspoons vanilla extract. Add mayonnaise mixture to flour mixture. Mix well to combine. Spread batter into prepared baking pan. Bake for 16 to 18 minutes or until a toothpick inserted in center comes out clean, being careful not to overbake. Place pan on a wire rack and allow to cool for 10 minutes. Punch holes in partially cooled cake with tines of a fork. Evenly drizzle chocolate syrup over partially cooled cake. In a large bowl, combine dry pudding mixes and ⅔ cup dry milk powder. Add 2¼ cups yogurt and remaining

1½ cups water. Mix well using a wire whisk. In another large bowl, combine remaining ¾ cup yogurt, remaining ⅔ cup dry milk powder, remaining 2 tablespoons Sugar Twin and remaining 2 teaspoons vanilla extract. Stir in Cool Whip Free. Cut cake into 16 pieces. To assemble, layer half of soaked cake pieces in bottom of glass trifle dish, cover with half of chocolate pudding mixture, sprinkle 2 tablespoons toffee chips over pudding, spoon half of yogurt mixture over top, then repeat layers. Cover and refrigerate for at least 1 hour.

Each serving equals:

HE: ½ Bread • ½ Skim Milk • ½ Slider • 16 Optional Calories

---

145 Calories • 1 gm Fat • 7 gm Protein • 27 gm Carbohydrate • 407 mg Sodium • 175 mg Calcium • 1 gm Fiber

---

DIABETIC: 1½ Starch/Carbohydrate • 1 Skim Milk

Magic Chocolate Cake with Sauce

Who really knows what goes on once the oven door is closed, after all? You mix up a few ingredients, turn up the heat, and if you've followed directions, you've got a wonderful treat in store. Don't wonder about the "tricks" that make this cake so tasty—just dive in!

● Serves 8

> 1½ cups all-purpose flour
> ¼ cup unsweetened cocoa
> 1½ teaspoons baking soda
> ¾ cup Sugar Twin or Sprinkle Sweet
> ¾ cup Kraft fat-free mayonnaise
> 1¾ cups water☆
> 2½ teaspoons vanilla extract☆
> 1 (4-serving) package JELL-O sugar-free chocolate cook-and-serve pudding mix
> ⅔ cup Carnation Nonfat Dry Milk Powder
> 1 tablespoon + 1 teaspoon reduced-calorie margarine

Preheat oven to 350 degrees. Spray a 9-by-9-inch cake pan with butter-flavored cooking spray. In a large bowl, combine flour, cocoa, baking soda, and Sugar Twin. Add mayonnaise, ¾ cup water, and 1½ teaspoons vanilla extract. Mix well to combine. Spread batter into prepared cake pan. Bake for 25 minutes or until toothpick inserted in center comes out clean, being careful not to overbake. Place cake pan on a wire rack and allow to cool. Meanwhile, in a medium saucepan, combine dry pudding mix, dry milk powder, remaining 1 cup water, remaining 1 teaspoon vanilla extract, and margarine. Cook over medium heat, stirring constantly with a wire whisk, until mixture thickens and starts to boil. Remove from heat. Cut partially cooled cake into 8 servings. When serving, place 1 piece of cake on a dessert plate and spoon about 2 tablespoons hot chocolate sauce over top.

HINTS: 1. Reheat leftover sauce in microwave.

           2. If you want to feel truly decadent, spoon 1 teaspoon chopped pecans and 1 tablespoon Cool Whip Lite over hot sauce.

Each serving equals:

HE: 1 Bread • ¼ Fat • ¼ Skim Milk • ½ Slider • 1 Optional Calorie

---

149 Calories • 1 gm Fat • 5 gm Protein • 30 gm Carbohydrate • 527 mg Sodium • 76 mg Calcium • 2 gm Fiber

---

DIABETIC: 2 Starch/Carbohydrate

Washington Cherry Cake ❄

If our country's first president needed a good reason to tackle that cherry tree, this recipe might have been enough! This creamy, fruity creation will win your family's vote for sure. ☻   Serves 8

> 1½ cups Bisquick Reduced Fat Baking Mix
> **2 recipes JO's Vanilla Milk Beverage Mix**
> 1 (4-serving) package JELL-O sugar-free instant vanilla
>   pudding mix
> ½ cup unsweetened apple juice
> 1 cup (one 8-ounce can) crushed pineapple, packed in
>   fruit juice, drained
> 2 tablespoons (½ ounce) chopped pecans
> 2 cups (one 16-ounce can) tart red cherries, packed in water,
>   drained

Preheat oven to 350 degrees. Spray a 9-by-9-inch cake pan with butter-flavored cooking spray. In a large bowl, combine baking mix, dry **JO's Vanilla Milk Beverage Mixes**, and dry pudding mix. Add apple juice and pineapple. Mix gently to combine. Stir in pecans and cherries. Evenly spread batter into prepared cake pan. Bake for 40 to 45 minutes or until a toothpick inserted in center comes out clean. Place cake pan on a wire rack and allow to cool completely. Cut into 8 servings.

HINT: Good served with 1 tablespoon Cool Whip Lite, but don't forget to count the few additional calories.

Each serving equals:

> HE: 1 Bread • ¾ Fruit • ¼ Fat • ¼ Skim Milk •
> 14 Optional Calories
>
> ---
>
> 118 Calories • 2 gm Fat • 3 gm Protein •
> 22 gm Carbohydrate • 262 mg Sodium •
> 86 mg Calcium • 1 gm Fiber
>
> ---
>
> DIABETIC: 1 Starch/Carbohydrate • ½ Fruit

Lemon Pudding Tarts

Here's a recipe that means you're always ready for unexpected company. Keep a package or two of graham cracker crusts in your pantry, quickly stir up the tart and tangy filling, and you're ready to welcome guests at any hour! Pam likes the classic taste and texture of this lovely dessert. ☉ Serves 6

> 1 (4-serving) package JELL-O sugar-free instant vanilla
>   pudding mix
> 1 (4-serving) package JELL-O sugar-free lemon gelatin
> ⅔ cup Carnation Nonfat Dry Milk Powder
> 1½ cups water
> ¾ cup Cool Whip Free
> 1 (6-single-serve) package Keebler graham cracker crusts
> 6 thin slices lemon (optional)

In a large bowl, combine dry pudding mix, dry gelatin, and dry milk powder. Add water. Mix well using a wire whisk. Blend in Cool Whip Free. Evenly spoon pudding mixture into crusts. Garnish each with a lemon slice. Refrigerate for at least 30 minutes.

Each serving equals:

HE: ⅔ Skim Milk • ½ Bread • ⅓ Skim Milk •
1 Slider • 8 Optional Calories

---

178 Calories • 6 gm Fat • 4 gm Protein •
27 gm Carbohydrate • 452 mg Sodium •
92 mg Calcium • 0 gm Fiber

DIABETIC: 1 Starch • 1 Fat • ½ Skim Milk
or 1½ Starch/Carbohydrate • 1 Fat

Rhubarb Spice Cake ❄

Cliff and I are both fans of rhubarb, those wonderful rosy stalks that inspire such scrumptious cakes and pies! This is a great choice for a charity bake sale or after-school snacks. It freezes beautifully too.

◐ Serves 8

> 1½ cups all-purpose flour
> 1 (4-serving) package JELL-O sugar-free instant vanilla
>   pudding mix
> ½ cup Sugar Twin or Sprinkle Sweet
> 1 teaspoon apple pie spice
> 1 teaspoon baking soda
> 2 cups finely chopped rhubarb
> ¼ cup (1 ounce) chopped walnuts
> **1 recipe JO's Sour Cream**
> 1 teaspoon cornstarch
> 1 egg or equivalent in egg substitute
> 1½ teaspoons vanilla extract
> ¼ cup water
> 2 tablespoons Brown Sugar Twin

Preheat oven to 350 degrees. Spray a 9-by-9-inch cake pan with butter-flavored cooking spray. In a large bowl, combine flour, dry pudding mix, Sugar Twin, apple pie spice, and baking soda. Stir in rhubarb and walnuts. In a small bowl, combine **JO's Sour Cream** and cornstarch. Add egg. Mix well to combine. Stir in vanilla extract and water. Add mixture to flour mixture. Mix gently just to combine. (Batter will be stiff.) Spread batter into prepared cake pan. Evenly sprinkle Brown Sugar Twin over top. Bake for 30 to 35 minutes or until a toothpick inserted in center comes out clean. Place cake pan on a wire rack and allow to cool. Cut into 8 servings.

HINTS: 1. Good warm or cold.
2. Also good topped with 1 tablespoon Cool Whip Lite, but don't forget to count the few additional calories.

Each serving equals:

HE: 1 Bread • ½ Vegetable • ¼ Protein • ¼ Fat • ¼ Skim Milk • ¼ Slider

---

155 Calories • 3 gm Fat • 6 gm Protein •
26 gm Carbohydrate • 364 mg Sodium •
113 mg Calcium • 1 gm Fiber

DIABETIC: 1½ Starch/Carbohydrate • ½ Fat

Jamaican Banana Spice Cake with Coconut Topping

This taste of the tropics will transform the cloudiest day into a sunny paradise—at least in your dining room! Rum, banana, coconut—how did I end up at this resort? ❤   Serves 8

1½ cups all-purpose flour

1 (4-serving) package JELL-O sugar-free instant
  banana pudding mix

¼ cup + 1 tablespoon Sugar Twin or Sprinkle Sweet☆

1 teaspoon baking powder

1 teaspoon baking soda

1½ teaspoons apple pie spice☆

⅔ cup (2 medium) mashed bananas

1 recipe JO's Sour Cream

1 teaspoon cornstarch

1 egg or equivalent in egg substitute

¼ cup water

1 tablespoon rum extract

2 tablespoons flaked coconut

Preheat oven to 350 degrees. Spray a 9-by-9-inch cake pan with butter-flavored cooking spray. In a large bowl, combine flour, dry pudding mix, ¼ cup Sugar Twin, baking powder, baking soda, and 1 teaspoon apple pie spice. In a medium bowl, combine bananas and **JO's Sour Cream**. Add cornstarch, egg, water, and rum extract. Mix well to combine. Gently stir banana mixture into flour mixture, just to combine. Spread batter into prepared loaf pan. In a small bowl, combine coconut, remaining 1 tablespoon Sugar Twin, and remaining ½ teaspoon apple pie spice. Evenly sprinkle topping mixture over batter. Bake for 20 to 25 minutes or until toothpick inserted in center comes out clean. Place pan on a wire rack and allow to cool. Cut into 8 servings.

Each serving equals:

HE: 1 Bread • ½ Fruit • ¼ Skim Milk • ¼ Slider • 6 Optional Calories

---

133 Calories • 1 gm Fat • 5 gm Protein • 26 gm Carbohydrate • 431 mg Sodium • 117 mg Calcium • 1 gm Fiber

---

DIABETIC: 1½ Starch/Carbohydrate • ½ Fruit

Apple Raisin Cream Treats

See if your family doesn't come running when the apples, raisins, and spices give off their sweet aroma as they cook! You may have to push them out of your kitchen so these pretty tarts can cool before serving.   ☻   Serves 6

> 1 (4-serving) package JELL-O sugar-free vanilla
>   cook-and-serve pudding mix
> 1 (4-serving) package JELL-O sugar-free lemon gelatin
> 1 cup water
> 2 cups (4 small) diced cooking apples
> 1/4 cup raisins
> 1 teaspoon apple pie spice
> **1 recipe JO's Deliteful Whipped Topping**
> 1 (6-single-serve) package Keebler graham cracker crusts

In a medium saucepan, combine dry pudding mix, dry gelatin, and water. Add apples, raisins, and apple pie spice. Mix well to combine. Cook over medium heat for 10 minutes or until mixture thickens and apples soften, stirring often. Remove from heat. Place pan on a wire rack and allow to cool completely. Add **JO's Deliteful Whipped Topping** to cooled apple mixture. Mix gently to combine. Evenly spoon mixture into crusts. (Tarts will be full.) Refrigerate for at least 30 minutes.

Each serving equals:

HE: 1 Fruit • 1/2 Bread • 1/3 Skim Milk • 1 Slider •
7 Optional Calories

226 Calories • 6 gm Fat • 5 gm Protein •
38 gm Carbohydrate • 312 mg Sodium •
108 mg Calcium • 1 gm Fiber

DIABETIC: 1 Fruit • 1 Starch • 1 Fat • 1/2 Skim Milk
or 2 Starch/Carbohydrate • 1 Fat • 1/2 Skim Milk

Frosted Raspberry Summer Pie ❄

Chocolate and raspberries go together like summer and sun! The rosy-tinted creamy topping, sprinkled with nuts and chips, looks irresistibly pretty. ☻   Serves 8

1½ cups fresh raspberries

1 (6-ounce) Keebler chocolate piecrust

1 (4-serving) package JELL-O sugar-free instant vanilla
     pudding mix

⅔ cup Carnation Nonfat Dry Milk Powder

1⅓ cups water

1 cup Cool Whip Free☆

½ teaspoon almond extract

3 to 4 drops red food coloring

1 tablespoon (¼ ounce) finely chopped almonds

1 tablespoon (¼ ounce) mini chocolate chips

Layer raspberries in bottom of piecrust. In a large bowl, combine dry pudding mix, dry milk powder, and water. Mix well using a wire whisk. Blend in ¼ cup Cool Whip Free. Spread pudding mixture evenly over raspberries. Refrigerate while preparing topping. In a small bowl, combine remaining ¾ cup Cool Whip Free, almond extract, and red food coloring. Evenly spread topping mixture over set filling. Sprinkle almonds and chocolate chips evenly over top. Refrigerate for at least 1 hour. Cut into 8 servings.

Each serving equals:

HE: ½ Bread • ¼ Fruit • ¼ Skim Milk • 1 Slider •
8 Optional Calories

174 Calories • 6 gm Fat • 3 gm Protein •
27 gm Carbohydrate • 301 mg Sodium •
77 mg Calcium • 1 gm Fiber

DIABETIC: 2 Starch/Carbohydrate • 1 Fat

Strawberry Chocolate Cream Pie ❄

Sometimes you just need a decadent dessert to celebrate a happy occasion. This luscious pie would be my choice (it's got strawberries, after all!). The drizzled chocolate syrup makes a great pie even better! ☻ Serves 8

> 2 cups sliced fresh strawberries
> 1 (6-ounce) Keebler chocolate piecrust
> 1 (4-serving) package JELL-O sugar-free instant chocolate
>   pudding mix
> ⅔ cup Carnation Nonfat Dry Milk Powder
> 1¼ cups water
> 1 cup Cool Whip Free☆
> 1 teaspoon almond extract☆
> 2 tablespoons (½ ounce) finely chopped slivered almonds
> 2 teaspoons Hershey's Lite Chocolate Syrup

Layer strawberries in bottom of piecrust. In a large bowl, combine dry pudding mix, dry milk powder, and water. Mix well using a wire whisk. Blend in ¼ cup Cool Whip Free and ½ teaspoon almond extract. Spread pudding mixture evenly over strawberries. Refrigerate while preparing topping. In a small bowl, combine remaining ¾ cup Cool Whip Free, remaining ½ teaspoon almond extract, and almonds. Spread topping mixture evenly over set filling. Drizzle chocolate syrup evenly over top. Refrigerate for at least 1 hour. Cut into 8 servings.

Each serving equals:

HE: ½ Bread • ¼ Fruit • ¼ Skim Milk • 1 Slider •
11 Optional Calories

---

182 Calories • 6 gm Fat • 4 gm Protein •
28 gm Carbohydrate • 302 mg Sodium •
79 mg Calcium • 1 gm Fiber

DIABETIC: 2 Starch/Carbohydrate • 1 Fat

Holiday Banana Nog Pie

Eggnog flavor in a scrumptious pie? It's an idea your family and friends will surely cheer. When you're busy before the holidays, you need a pie that whips up this quickly and tastes oh-so-good! (Son-in-law John calls this "lethal," a dessert "to die for," and he's been stirring it up for company ever since he tasted it!)

❂ Serves 8

> 2 cups (2 medium) diced bananas
> 1 (6-ounce) Keebler shortbread piecrust
> 1 (4-serving) package JELL-O sugar-free instant vanilla
>   pudding mix
> ⅔ cup Carnation Nonfat Dry Milk Powder
> 1¼ cups water
> ¾ cup Cool Whip Free☆
> 1 teaspoon rum extract
> ¼ teaspoon ground nutmeg

Layer bananas in bottom of piecrust. In a large bowl, combine dry pudding mix, dry milk powder, and water. Mix well using a wire whisk. Blend in ¼ cup Cool Whip Free, rum extract, and nutmeg. Spread pudding mixture evenly over bananas. Refrigerate for 10 minutes. Evenly spread remaining ½ cup Cool Whip Free over set filling and lightly sprinkle additional nutmeg over top. Refrigerate for at least 1 hour. Cut into 8 servings.

HINT: To prevent bananas from turning brown, mix with 1 teaspoon lemon juice or sprinkle with Fruit Fresh.

Each serving equals:

HE: ½ Bread • ½ Fruit • ¼ Skim Milk •
¾ Slider • 4 Optional Calories

178 Calories • 6 gm Fat • 3 gm Protein •
28 gm Carbohydrate • 331 mg Sodium •
71 mg Calcium • 1 gm Fiber

DIABETIC: 1 Starch • 1 Fruit • 1 Fat
or 2 Starch/Carbohydrate • 1 Fat

New England
Raisin-Butterscotch Pie

Sometimes I wonder how many ways those clever New Englanders find to use their local delicacy, maple syrup. I've come up with more than a few myself out here in Iowa! My daughter Becky loves butterscotch pies, and she gave this one a perfect "four-raisin" rating!　　●　　Serves 8

> 1 (4-serving) package JELL-O sugar-free instant butterscotch pudding mix
> ⅔ cup Carnation Nonfat Dry Milk Powder
> ¾ cup water
> ½ cup + 2 tablespoons Cary's Sugar Free Maple Syrup☆
> 1 cup raisins
> ¾ cup Cool Whip Free☆
> 1 (6-ounce) Keebler graham cracker piecrust
> 2 tablespoons (½ ounce) chopped pecans

In a large bowl, combine dry pudding mix, dry milk powder, water, and ½ cup maple syrup. Mix well using a wire whisk. Blend in raisins and ¼ cup Cool Whip Free. Spread mixture evenly into piecrust. Refrigerate while preparing topping. In a small bowl, gently combine remaining ½ cup Cool Whip Free and remaining 2 tablespoons maple syrup. Spread topping mixture evenly over set filling. Evenly sprinkle pecans over top. Refrigerate for at least 1 hour. Cut into 8 servings.

HINT:　To plump up raisins without "cooking," place in a glass measuring cup and microwave on HIGH for 45 seconds.

Each serving equals:

> HE: 1 Fruit • ½ Bread • ¼ Skim Milk • ¼ Fat • 1 Slider • 6 Optional Calories

> 230 Calories • 6 gm Fat • 4 gm Protein • 40 gm Carbohydrate • 384 mg Sodium • 78 mg Calcium • 1 gm Fiber

> DIABETIC: 1½ Starch/Carbohydrate • 1 Fruit • 1 Fat

Chocolate "Bon Bon" Pie

I love the idea that "bon-bon," the French word for a sweet treat, translates as "good-good"! This chocolate raspberry pie is so good, it deserves to be complimented twice—it's got two chocolate ingredients *and* two raspberry ones!　　●　　Serves 8

> 2 (4-serving) packages JELL-O sugar-free instant chocolate pudding mix
> 1 (4-serving) package JELL-O sugar-free raspberry gelatin
> 1⅓ cups Carnation Nonfat Dry Milk Powder
> 2¼ cups water
> 1 (6-ounce) Keebler chocolate piecrust
> ½ cup raspberry spreadable fruit spread
> ¾ cup Cool Whip Free

In a large bowl, combine dry pudding mix, dry gelatin, and dry milk powder. Add water. Mix well using a wire whisk. Spread pudding mixture into piecrust. Refrigerate while preparing topping. In a small bowl, stir fruit spread until soft. Add Cool Whip Free. Mix gently to combine. Spread topping mixture evenly over set filling. Refrigerate for at least 1 hour. Cut into 8 servings.

HINT: Any compatible gelatin and spreadable fruit combination may be used.

Each serving equals:

HE: 1 Fruit • ½ Bread • ½ Skim Milk • 1 Slider • 16 Optional Calories

238 Calories • 6 gm Fat • 6 gm Protein • 40 gm Carbohydrate • 519 mg Sodium • 139 mg Calcium • 1 gm Fiber

DIABETIC: 1 Starch/Carbohydrate • 1 Fruit • 1 Fat • ½ Skim Milk

German Sweet Chocolate Pie ❄

Here's some good news—you don't have to fly all the way to Germany to enjoy this chocolate, coconut, and pecan combo! The flavors are so rich, no one will believe they're enjoying a "healthy" pie. ♥ Serves 8

2 (4-serving) packages JELL-O sugar-free chocolate cook-and-serve pudding mix
1⅓ cups Carnation Nonfat Dry Milk Powder
3 cups water
1 teaspoon vanilla extract
1 teaspoon coconut extract
1 (6-ounce) Keebler chocolate piecrust
6 tablespoons purchased graham cracker crumbs or
 6 (2½-inch) graham crackers, made into fine crumbs☆
2 tablespoons flaked coconut☆
2 tablespoons (½ ounce) chopped pecans☆

Preheat oven to 350 degrees. In a medium saucepan, combine dry pudding mixes, dry milk powder, and water. Mix well to combine. Cook over medium heat for 5 minutes or until mixture thickens and starts to boil, stirring constantly. Remove from heat. Stir in vanilla and coconut extracts. Pour ⅓ of pudding mixture into piecrust. Evenly sprinkle ½ of cracker crumbs, 1 teaspoon coconut, and 1 teaspoon pecans over top. Gently spoon remaining pudding over top of crumbs. Evenly sprinkle remaining cracker crumbs, remaining coconut, and pecans evenly over the top. Bake for 30 minutes. Place pie plate on a wire rack and allow to cool completely. Refrigerate for at least 2 hours. Cut into 8 servings.

Each serving equals:

HE: ¾ Bread • ½ Skim Milk • ¼ Fat • ¾ Slider • 19 Optional Calories

215 Calories • 7 gm Fat • 6 gm Protein • 32 gm Carbohydrate • 309 mg Sodium • 141 mg Calcium • 1 gm Fiber

DIABETIC: 2 Starch/Carbohydrate • 1 Fat

Mercy Memories Pumpkin Pie

This pie was originally created to honor a hospital where I did a cooking demonstration, but it's the cozy memories and wonderful taste your family will recall after feasting on this luscious dessert!

○ Serves 8

> ⅔ cup Carnation Nonfat Dry Milk Powder
>
> ½ cup water
>
> 2 cups (one 16-ounce can) pumpkin
>
> 1 (4-serving) package JELL-O sugar-free instant butterscotch pudding mix
>
> 1 teaspoon pumpkin pie spice
>
> ½ cup Cool Whip Free
>
> 1 (6-ounce) Keebler graham cracker piecrust
>
> 2 tablespoons (½ ounce) chopped pecans
>
> ½ cup Cary's Sugar Free Maple Syrup

In a large bowl, combine dry milk powder and water. Add pumpkin, dry pudding mix, and pumpkin pie spice. Blend in Cool Whip Free. Spread mixture evenly into piecrust. Evenly sprinkle pecans over top of pie. Refrigerate for at least 1 hour. Cut into 8 servings. Just before serving, drizzle 1 tablespoon maple syrup over top of each piece.

Each serving equals:

HE: ½ Bread • ½ Vegetable • ¼ Skim Milk • ¼ Fat • 1 Slider

186 Calories • 6 gm Fat • 4 gm Protein • 29 gm Carbohydrate • 347 mg Sodium • 89 mg Calcium • 2 gm Fiber

DIABETIC: 2 Starch/Carbohydrate • 1 Fat

Magical Pumpkin Pie

In our family, pumpkin pie MUST be on the menu for both Thanksgiving and Christmas, so I've created many different versions over the years. This one deserves the name "magical" because its few ingredients transform so easily and deliciously into a holiday treasure. ☻ Serves 8

> 1 Pillsbury refrigerated unbaked 9-inch piecrust
>
> 2 cups (one 16-ounce can) pumpkin
>
> **1 recipe JO's Sweetened Condensed Milk**
>
> 1 egg or equivalent in egg substitute
>
> 1½ teaspoons pumpkin pie spice

Preheat oven to 375 degrees. Place piecrust in a 9-inch pie plate and flute edges. In a large bowl, combine pumpkin, **JO's Sweetened Condensed Milk**, egg, and pumpkin pie spice. Pour pumpkin mixture into prepared piecrust. Bake for 50 to 55 minutes or until a knife inserted near center comes out clean. Place pie plate on a wire rack and allow to cool for 15 minutes. Refrigerate at least 1 hour. Cut into 8 servings.

Each serving equals:

HE: ½ Bread • ½ Vegetable • ½ Skim Milk • ¾ Slider • 4 Optional Calories

192 Calories • 8 gm Fat • 6 gm Protein • 24 gm Carbohydrate • 173 mg Sodium • 161 mg Calcium • 2 gm Fiber

DIABETIC: 1 Starch • 1 Fat • ½ Skim Milk

Fudgy Oatmeal Pie

Imagine eating a giant oatmeal peanut butter cookie—and you'll have some idea how this unique dessert will taste! It's a real Healthy Exchanges original, and I hope you'll enjoy it as much as Cliff and I do. ☻ Serves 8

1 Pillsbury refrigerated unbaked 9-inch piecrust

1 cup (3 ounces) quick oats

1⅓ cups Carnation Nonfat Dry Milk Powder

3 tablespoons unsweetened cocoa

1 teaspoon baking powder

½ cup Sugar Twin or Sprinkle Sweet

1¼ cups water

2 tablespoons Peter Pan reduced-fat peanut butter

1 teaspoon vanilla extract

Preheat oven to 400 degrees. Place piecrust in a 9-inch pie plate. Flute edges. In a medium bowl, combine oats, dry milk powder, cocoa, baking powder, and Sugar Twin. Add water, peanut butter, and vanilla extract. Mix well to combine. Pour mixture into piecrust. Bake for 10 minutes. Lower heat to 350 degrees. Continue baking for 20 minutes. Place pie plate on a wire rack and allow to cool. Cut into 8 servings.

HINT: Good served with 1 tablespoon Cool Whip Lite, but don't forget to count the few additional calories.

Each serving equals:

HE: 1 Bread • ½ Skim Milk • ¼ Fat •
¼ Protein • ¾ Slider • 2 Optional Calories

225 Calories • 9 gm Fat • 7 gm Protein •
29 gm Carbohydrate • 243 mg Sodium •
181 mg Calcium • 2 gm Fiber

DIABETIC: 1½ Starch • 1 Fat • ½ Meat •
½ Skim Milk

Broadcast House
Cherry Blossom Cheesecake ❄

This delicious cherry cheesecake was "born" in a television studio in Washington, D.C., where each spring the cherry blossoms turn our nation's capital into a fragrant and exquisitely beautiful city of dreams. ☕ Serves 8

> 2 (8-ounce) packages Philadelphia fat-free cream cheese
> 1 (4-serving) package JELL-O sugar-free instant vanilla
>   pudding mix
> 2/3 cup Carnation Nonfat Dry Milk Powder
> 1 cup Diet 7-UP
> 1/4 cup Cool Whip Free
> 1 1/2 teaspoons coconut extract☆
> 1 (6-ounce) Keebler shortbread piecrust
> 1 (4-serving) package JELL-O sugar-free cherry gelatin
> 1 (4-serving) package JELL-O sugar-free vanilla
>   cook-and-serve pudding mix
> 2 cups (one 16-ounce can) tart red cherries, packed in water,
>   drained, and 1/3 cup liquid reserved
> 1 cup water
> 2 tablespoons flaked coconut

In a large bowl, stir cream cheese with a spoon until soft. Add dry instant pudding mix, dry milk powder, and Diet 7-UP. Mix well using a wire whisk. Blend in Cool Whip Free and 1 teaspoon coconut extract. Spread mixture evenly into piecrust. Refrigerate. Meanwhile in a medium saucepan, combine dry gelatin and dry cook-and-serve pudding mix. Stir in reserved cherry liquid, water, and cherries. Cook over medium heat for 6 to 8 minutes or until mixture thickens and starts to boil, stirring often and being careful not to crush cherries. Remove from heat. Stir in remaining 1/2 teaspoon coconut extract. Place saucepan on a wire rack and allow to cool for 10 minutes. Evenly spoon partially cooled cherry mixture over top of filling. Refrigerate for at least 1 hour. Just before serving, sprinkle coconut evenly over top. Cut into 8 servings.

Each serving equals:

HE: 1 Protein • ½ Bread • ½ Fruit • ¼ Skim Milk • 1 Slider • 5 Optional Calories

225 Calories • 5 gm Fat • 12 gm Protein • 33 gm Carbohydrate • 727 mg Sodium • 76 mg Calcium • 1 gm Fiber

DIABETIC: 1 Meat • 1 Starch/Carbohydrate • 1 Fruit • ½ Fat

Banana Split Cheesecake

Every year I create at least one new "banana split" dessert that's as fun to prepare as it is to eat! Think of this one as a sundae onto which you get to pile the goodies high.　　❂　　Serves 8

> 2 (8-ounce) packages Philadelphia fat-free cream cheese
> 1 (4-serving) package JELL-O sugar-free instant banana
> 　　pudding mix
> ⅔ cup Carnation Nonfat Dry Milk Powder
> 1 cup (one 8-ounce can) crushed pineapple, packed in fruit juice,
> 　　undrained
> 2 cups chopped fresh strawberries
> 1 cup (one medium) diced banana
> 1 (6-ounce) Keebler shortbread piecrust
> ¾ cup Cool Whip Free
> 1 teaspoon coconut extract
> 2 tablespoons (½ ounce) chopped pecans
> 1 tablespoon (¼ ounce) mini chocolate chips
> 1 tablespoon flaked coconut
> 4 maraschino cherries, halved

In a large bowl, stir cream cheese with a spoon until soft. Add dry pudding mix, dry milk powder, and undrained pineapple. Mix well using a wire whisk. Gently fold in strawberries and banana. Spread mixture evenly into piecrust. Refrigerate while preparing topping. In a small bowl, combine Cool Whip Free and coconut extract. Spread topping mixture evenly over top of filling. Evenly sprinkle pecans, chocolate chips, and coconut over top. Garnish with cherry halves. Refrigerate for at least 1 hour. Cut into 8 servings.

HINT:　To prevent banana from turning brown, mix with 1 teaspoon lemon juice or sprinkle with Fruit Fresh.

Each serving equals:

HE: 1 Protein • ¾ Fruit • ½ Bread • ¼ Fat •
¼ Skim Milk • 1 Slider • 7 Optional Calories

262 Calories • 6 gm Fat • 12 gm Protein •
40 gm Carbohydrate • 642 mg Sodium •
81 mg Calcium • 2 gm Fiber

DIABETIC: 1½ Starch/Carbohydrate • 1 Meat •
1 Fruit • 1 Fat

Magic Mocha Cheesecake

Sweetened condensed milk is a bit of baking magic, and my low-fat, low-sugar version makes this mocha treat a "miraculous" munchie! ☽ Serves 8

1 tablespoon instant coffee crystals
¼ cup hot water
1 (8-ounce) package Philadelphia fat-free cream cheese
1 recipe JO's Sweetened Condensed Milk
1 (4-serving) package JELL-O sugar-free instant chocolate
    pudding mix
1 cup Cool Whip Free
1 (6-ounce) Keebler chocolate piecrust
2 (2½-inch) chocolate graham crackers,
    made into fine crumbs

In a small bowl, combine coffee crystals and hot water. Set aside. In a large bowl, stir cream cheese with a spoon until soft. Add **JO's Sweetened Condensed Milk**, dry pudding mix, and partially cooled coffee mixture. Mix well using a wire whisk. Blend in Cool Whip Free. Evenly spread mixture into piecrust. Sprinkle chocolate crumbs evenly over the top. Cover and refrigerate for at least 2 hours. Cut into 8 servings.

Each serving equals:

HE: ½ Protein • ½ Bread • ½ Skim Milk • 1 Slider •
12 Optional Calories

205 Calories • 5 gm Fat • 10 gm Protein •
30 gm Carbohydrate • 513 mg Sodium •
140 mg Calcium • 1 gm Fiber

DIABETIC: 1½ Starch • ½ Meat • ½ Skim Milk
or 2 Starch/Carbohydrate • ½ Meat

This and That

Pineapple-Orange Fruit Dip

Think of this as Healthy Exchanges fruit fondue, and dip away! It's a fun idea for a party anytime at all.   ☺   Serves 6 (⅓ cup)

¾ cup Yoplait plain fat-free yogurt

⅔ cup Carnation Nonfat Dry Milk Powder

1 (4-serving) package JELL-O sugar-free orange gelatin

1 cup (one 8-ounce can) crushed pineapple, packed in fruit juice,
    well drained

⅓ cup Cool Whip Free

1 teaspoon coconut extract

In a medium bowl, combine yogurt and dry milk powder. Add dry gelatin. Mix well to combine. Stir in pineapple, Cool Whip Free, and coconut extract. Cover and refrigerate for at least 1 hour. Gently stir again just before serving.

HINT: Wonderful served with fresh pineapple chunks and whole fresh strawberries.

Each serving equals:

HE: ½ Skim Milk • ⅓ Fruit • 13 Optional Calories

80 Calories • 0 gm Fat • 5 gm Protein •
15 gm Carbohydrate • 99 mg Sodium •
154 mg Calcium • 0 gm Fiber

DIABETIC: 1 Fruit *or* 1 Skim Milk

South of the Border Shrimp Dip

There are lots of party possibilities in this spicy seafood combo. I like to use it on fresh vegetables or crackers, or as a tangy topping for a steaming baked potato.   ♥   Serves 6 (⅓ cup)

> *1 recipe JO's Sour Cream*
> *2 teaspoons prepared horseradish sauce*
> *1 teaspoon dried parsley flakes*
> *¾ cup chunky salsa (mild, medium, or hot)*
> *1 (4.25-ounce can) small shrimp, rinsed and drained*

In a medium bowl, combine **JO's Sour Cream**, horseradish sauce, and parsley flakes. Add salsa and shrimp. Mix well to combine. Cover and refrigerate for at least 30 minutes. Gently stir again just before serving.

HINT: Drain excess liquid from salsa, if not using extra-thick type.

Each serving equals:

HE: ⅔ Protein • ⅓ Skim Milk • ¼ Vegetable •
4 Optional Calories

---

52 Calories • 0 gm Fat • 8 gm Protein •
5 gm Carbohydrate • 188 mg Sodium •
156 mg Calcium • 0 gm Fiber

DIABETIC: ½ Meat • ½ Starch/Carbohydrate

Pita Eggs Olé

Fed up with the same old breakfasts week after week? Take a chance on this spicy ham-and-egg omelet served in a pita, and wake up your day in a lively new way. You can even eat this one on the run! ☻ Serves 4

> 1 full cup (6 ounces) finely diced Dubuque 97% fat-free
>   ham or any extra-lean ham
> 4 eggs or equivalent in egg substitute
> ½ cup chunky salsa (mild, medium, or hot)
> ¼ teaspoon lemon pepper
> 1 teaspoon dried parsley flakes
> ⅓ cup (1½ ounces) shredded Kraft reduced-fat
>   Cheddar cheese
> 2 pita rounds, halved

In a large skillet sprayed with olive oil–flavored cooking spray, brown ham. In a medium bowl, beat eggs with a fork. Stir in salsa, lemon pepper, and parsley flakes. Add egg mixture to ham. Mix well to combine. Continue cooking until eggs start to set, stirring occasionally. Stir in Cheddar cheese. Lower heat, cover, and simmer for 3 to 4 minutes or until cheese is melted and eggs are set. Fluff gently with a fork. Evenly spoon about ½ cup egg mixture into each pita half. Serve at once.

Each serving equals:

HE: 2½ Protein (1 limited) • 1 Bread • ¼ Vegetable

226 Calories • 8 gm Fat • 18 gm Protein •
20 gm Carbohydrate • 603 mg Sodium •
160 mg Calcium • 1 gm Fiber

DIABETIC: 2½ Meat • 1 Starch

Grande Egg Skillet

I'm not sure how to describe this, though it's sort of a tangy, cheese French toast! Oh, just taste it and see if you don't feel energized and excited about the day's events.   ☻   Serves 4

> 4 eggs or equivalent in egg substitute
> ½ teaspoon lemon pepper
> 4 slices reduced-calorie bread, toasted and cut into cubes
> ⅓ cup (1½ ounces) shredded Kraft reduced-fat
>   Cheddar cheese
> ½ cup chunky salsa (mild, medium, or hot)
> ¼ cup Land O Lakes no-fat sour cream

In a medium bowl, mix eggs and lemon pepper with a wire whisk until fluffy. Place toast pieces in a large skillet sprayed with butter-flavored cooking spray. Pour eggs over top. Cook over medium heat until eggs begin to set, stirring occasionally. Sprinkle Cheddar cheese evenly over eggs. Cover and remove from heat. Let set for 2 to 3 minutes or until cheese melts. Cut into 4 wedges. For each serving, place 1 wedge on a serving plate, top each piece with 2 tablespoons salsa and garnish with 1 tablespoon sour cream.

Each serving equals:

> 1½ Protein (1 limited) • ½ Bread • ¼ Vegetable •
> 15 Optional Calories
> ───────────────────────────────
> 163 Calories • 7 gm Fat • 12 gm Protein •
> 13 gm Carbohydrate • 380 mg Sodium •
> 168 mg Calcium • 2 gm Fiber
> ───────────────────────────────
> DIABETIC: 1½ Meat • 1 Starch

Eggs à la King

Want to start the day with a calcium booster? This creamy egg delight packs a real nutritional wallop—but it's the fiesta of flavors that will really make your morning.   ☻   Serves 4

2 cups skim milk

3 tablespoons all-purpose flour

2 tablespoons Heinz Light Harvest or Healthy Choice Ketchup

4 hard-boiled eggs, sliced

½ cup frozen peas

2 tablespoons chopped canned pimientos

½ cup (one 2.5-ounce jar) sliced mushrooms, drained

¼ teaspoon black pepper

8 slices reduced-calorie bread, toasted

In a covered jar, combine skim milk and flour. Shake well to blend. Pour mixture into a large skillet sprayed with butter-flavored cooking spray. Cook over medium heat for 5 minutes or until mixture thickens, stirring constantly. Stir in ketchup. Add eggs, peas, pimientos, mushrooms, and black pepper. Mix well to combine. Lower heat and simmer for 5 minutes or until mixture is heated through, stirring often. For each serving, place 2 slices of toast on a serving plate and spoon about ¾ cup egg mixture over top.

Each serving equals:

HE: 1½ Bread • 1 Protein (limited) • ¼ Skim Milk • ¼ Vegetable

---

250 Calories • 6 gm Fat • 18 gm Protein • 31 gm Carbohydrate • 391 mg Sodium • 220 mg Calcium • 2 gm Fiber

---

DIABETIC: 1½ Starch • 1 Meat • ½ Skim Milk

Egg-Cheese Sandwiches

Time for a new take on egg salad, this one blended with shredded cheese for a fun new taste sensation that'll chase your blues away! My junior taste-tester, Zach, loves creamy sandwich fillings, so this one got his "thumbs-up"!   ◐   Serves 4

¼ cup Kraft fat-free mayonnaise

Sugar substitute to equal 2 teaspoons sugar

1 teaspoon prepared mustard

1 teaspoon white vinegar

2 teaspoons sweet pickle relish

¼ teaspoon black pepper

3 hard-boiled eggs, chopped

¾ cup (3 ounces) shredded Kraft reduced-fat
    Cheddar cheese

4 reduced-calorie hamburger buns

In a medium bowl, combine mayonnaise, sugar substitute, mustard, vinegar, pickle relish, and black pepper. Add eggs and Cheddar cheese. Mix gently to combine. For each sandwich, spoon about ⅓ cup mixture between a bun. Serve at once or cover and refrigerate until ready to serve.

Each serving equals:

HE: 1¾ Protein (¾ limited) • 1 Bread •
13 Optional Calories

200 Calories • 8 gm Fat • 13 gm Protein •
19 gm Carbohydrate • 511 mg Sodium •
163 mg Calcium • 1 gm Fiber

DIABETIC: 1½ Meat • 1 Starch

Deli Bagelwich

Stumped for what to put on your bagel for brunch Sunday morning? Cliff just loves this colorful, tasty blend of flavors.

⏺ Serves 4

> ¼ cup Land O Lakes no-fat sour cream
>
> 1 teaspoon Dijon mustard
>
> 1 teaspoon dried parsley flakes
>
> 1 cup shredded carrots
>
> 4 (¾-ounce) slices Kraft reduced-fat Cheddar cheese
>
> 4 small bagels, sliced in half
>
> ½ cup shredded lettuce
>
> ½ cup tomato slices

In a medium bowl, combine sour cream, mustard, and parsley flakes. Stir in carrots. Set aside. For each sandwich, place 1 slice of Cheddar cheese on a bagel bottom, layer 2 tablespoons lettuce and about 2 slices tomato over cheese, spread about ¼ cup carrot mixture over tomato and arrange top half of bagel over top. Serve at once or cover and refrigerate until ready to serve.

HINT: Try this with onion bagels for a real tangy treat!

Each serving equals:

HE: 2 Bread • 1 Protein • 1 Vegetable •
15 Optional Calories

---

227 Calories • 4 gm Fat • 12 gm Protein •
38 gm Carbohydrate • 686 mg Sodium •
198 mg Calcium • 2 gm Fiber

DIABETIC: 2 Starch • 1 Meat • 1 Vegetable

Baked Raisin Pancake Squares ❄

Instead of waffles or traditional pancakes when your grandbabies come to visit, here's a kid-pleaser (tested and approved by Josh) that couldn't be easier to stir up and bake when they tug you out of bed! Trust me, it's a husband-pleaser too. This makes a dense pancake that seems to be a favorite in the Midwest. For a fluffier version, substitute regular skim milk for the evaporated kind. The fluffier version doesn't always freeze as well, however. ☺ Serves 8

> 2 cups Bisquick Reduced Fat Baking Mix
> 1/3 cup Carnation Nonfat Dry Milk Powder
> 1 teaspoon apple pie spice
> 1/2 cup raisins
> 1 1/2 cups (one 12-fluid-ounce can) Carnation Evaporated
>   Skim Milk
> 2 eggs or equivalent in egg substitute

Preheat oven to 400 degrees. Spray a rimmed 10-by-15-inch cookie sheet with a deep lip with butter-flavored cooking spray. In a large bowl, combine baking mix, dry milk powder, apple pie spice, and raisins. Add evaporated skim milk and eggs. Mix gently to combine. Pour mixture into prepared cookie sheet. Bake for 12 minutes or until center tests done. Place cookie sheet on a wire rack and let set for 5 minutes. Cut into 8 servings.

HINTS: 1. Good served with reduced-calorie maple syrup. If using, count optional calories accordingly.
2. Freeze leftovers in individual servings and thaw in microwave just before serving.

Each serving equals:

HE: 1 1/3 Bread • 1/2 Skim Milk • 1/2 Fruit •
1/4 Protein (limited)

203 Calories • 3 gm Fat • 9 gm Protein •
35 gm Carbohydrate • 436 mg Sodium •
208 mg Calcium • 1 gm Fiber

DIABETIC: 1 1/2 Starch • 1/2 Fruit • 1/2 Skim Milk
or 2 Starch/Carbohydrate • 1/2 Skim Milk

Pumpkin Muffins

Pumpkin is a great ingredient to bake with because it delivers flavor, color, and the moisture you need to produce a magnificent muffin! Be prepared for an irresistible aroma to emerge from the oven when you check on these! ☻ Serves 16 (1 each)

½ cup (1.5 ounces) quick oats☆
¾ cup all-purpose flour
1⅓ cups Carnation Nonfat Dry Milk Powder
⅓ cup Sugar Twin or Sprinkle Sweet
1 tablespoon baking soda
1 tablespoon pumpkin pie spice
½ cup (2 ounces) chopped walnuts
½ cup raisins
2 cups (one 16-ounce can) pumpkin
2 eggs or equivalent in egg substitute
1 tablespoon vanilla extract

Preheat oven to 350 degrees. Spray muffin pans with butter-flavored cooking spray or line with paper liners. In a large skillet sprayed with butter-flavored cooking spray, lightly brown oats. In a large bowl, combine flour, dry milk powder, Sugar Twin, baking soda, pumpkin pie spice, walnuts, and raisins. Stir in ¼ cup oats. In a medium bowl, combine pumpkin, eggs, and vanilla extract. Add pumpkin mixture to flour mixture. Stir just until moist. Evenly divide batter among 16 prepared muffin cups. Sprinkle about ¾ teaspoon browned oats evenly over top of each muffin. Bake for 20 minutes or until a toothpick inserted in center comes out clean. Place muffin pans on a wire rack and let set for 5 minutes. Remove muffins from pans and continue cooling on wire rack.

HINT: Fill unused muffin wells with water. It protects the muffin tin and ensures even baking.

Each serving equals:

HE: ⅓ Bread • ¼ Skim Milk • ¼ Fat • ¼ Protein •
¼ Fruit • 6 Optional Calories

111 Calories • 3 gm Fat • 5 gm Protein •
16 gm Carbohydrate • 278 mg Sodium •
91 mg Calcium • 2 gm Fiber

DIABETIC: 1 Starch • ½ Fat

Banana Cherry Coffee Cake ❄

Coffee cake isn't a no-no when it's prepared the Healthy Exchanges Way. I've whisked out excess fat and sugar, but stirred in so many delicious ingredients you'll never miss 'em! ☻ Serves 8

> 1½ cups Bisquick Reduced Fat Baking Mix
> 1 (4-serving) package JELL-O sugar-free instant banana
>   pudding mix
> ⅔ cup Carnation Nonfat Dry Milk Powder
> 1½ cups fresh cherries, pitted and well drained
> ¼ cup (1 ounce) chopped walnuts
> ⅔ cup (2 ripe medium) mashed bananas
> ½ cup unsweetened applesauce
> ½ cup water
> 1 egg, beaten, or equivalent in egg substitute
> ½ teaspoon ground cinnamon
> 2 tablespoons Sugar Twin or Sprinkle Sweet

Preheat oven to 350 degrees. Spray an 8-by-8-inch baking dish with butter-flavored cooking spray. In a large bowl, combine baking mix, dry pudding mix, and dry milk powder. Stir in cherries and walnuts. In a medium bowl, combine mashed bananas, applesauce, water, and egg. Add banana mixture to baking mix mixture. Mix well to combine. Spread batter in prepared baking dish. In a small saucer, combine cinnamon and Sugar Twin. Evenly sprinkle cinnamon mixture over top of batter. Bake for 30 to 40 minutes or until a toothpick inserted in center comes out clean. Place baking dish on a wire rack and allow to cool. Cut into 8 servings.

HINT: Unsweetened frozen cherries, thawed, or canned cherries packed in water may be used. Just be sure to drain either one well before using.

Each serving equals:

HE: 1 Bread • 1 Fruit • ¼ Skim Milk •
¼ Protein • ¼ Fat • 14 Optional Calories

180 Calories • 4 gm Fat • 5 gm Protein •
31 gm Carbohydrate • 474 mg Sodium •
102 mg Calcium • 1 gm Fiber

DIABETIC: 1 Fruit • 1 Starch • ½ Fat

Pineapple "Buttermilk" Coffee Cake

My JO's dairy mixes were created to keep you from having to rush to the store when you need an ingredient you don't often use, like buttermilk. This cake is lovely for a weekend brunch, but it's also a wonderful treat to enjoy just any old day. Quick to mix, quick to fix—and quick to satisfy your tummy! ☻ Serves 8

1½ cups Bisquick Reduced Fat Baking Mix
1 (4-serving) package JELL-O sugar-free instant vanilla
    pudding mix
¼ cup (1 ounce) chopped walnuts
1 cup (one 8-ounce can) crushed pineapple, packed in
    fruit juice, undrained
1 egg or equivalent in egg substitute
1 teaspoon vanilla extract
1 recipe JO's Buttermilk
½ teaspoon ground cinnamon
2 tablespoons Sugar Twin or Sprinkle Sweet

Preheat oven to 375 degrees. Spray an 8-by-8-inch baking dish with butter-flavored cooking spray. In a large bowl, combine baking mix, dry pudding mix, and walnuts. In a medium bowl, combine undrained pineapple, egg, and vanilla extract. Stir in **JO's Buttermilk**. Add pineapple mixture to baking mix mixture. Mix well to combine. Spread batter into prepared baking dish. In a small bowl, combine cinnamon and Sugar Twin. Sprinkle cinnamon mixture evenly over top of batter. Bake for 18 to 22 minutes or until a toothpick inserted in center comes out clean. Place baking dish on a wire rack and allow to cool. Cut into 8 servings.

Each serving equals:

HE: 1 Bread • ¼ Skim Milk • ¼ Fat • ¼ Protein •
¼ Fruit • 14 Optional Calories

168 Calories • 4 gm Fat • 5 gm Protein •
28 gm Carbohydrate • 466 mg Sodium •
100 mg Calcium • 0 gm Fiber

DIABETIC: 2 Starch

German Pancakes

If a European vacation isn't in the cards, why not take a quick culinary journey and serve these delicate and spicy circles? Remember to keep the pancakes thin—they're best that way!

☺ Serves 8

> 2 cups skim milk
> 2 eggs or equivalent in egg substitute
> 1½ cups all-purpose flour
> ¼ cup + 1 teaspoon Sugar Twin or Sprinkle Sweet☆
> 1 tablespoon ground cinnamon

In a blender container, combine skim milk, eggs, flour, and 1 teaspoon Sugar Twin. Cover and process on BLEND for 30 seconds. Using ¼ cup measure as a guide, pour batter onto a griddle or large skillet sprayed with butter-flavored cooking spray to form 8 pancakes. Gently flatten to form very thin pancakes. Lightly brown pancakes on both sides. In a small bowl, combine remaining ¼ cup Sugar Twin and cinnamon. Evenly sprinkle cinnamon mixture over warm pancakes when serving.

HINT: Also good served with warm applesauce spooned over top.

Each serving equals:

HE: 1 Bread • ¼ Skim Milk • ¼ Protein (limited) • 3 Optional Calories

121 Calories • 1 gm Fat • 6 gm Protein • 22 gm Carbohydrate • 181 mg Sodium • 95 mg Calcium • 1 gm Fiber

DIABETIC: 1½ Starch

Golden Gate Shakes

Instead of a costly health food store shake, why not blend up your own fruit-filled, good-for-you snack? I first tasted a shake like this in San Francisco, hence the name. The frozen berries take the place of ice cubes to make this extra frothy.

● Serves 4 (1 cup)

3 cups skim milk
1 (4-serving) package JELL-O sugar-free instant banana
    pudding mix
1 cup (one medium) sliced banana
3⁄4 cup frozen unsweetened blueberries

In a blender container, combine skim milk, dry pudding mix, and banana. Cover and process on BLEND for 15 seconds. Add frozen blueberries. Re-cover and process on HIGH for 10 seconds or until mixture is smooth. Serve at once.

Each serving equals:

HE: 3⁄4 Skim Milk • 3⁄4 Fruit • 1⁄4 Slider •
5 Optional Calories

136 Calories • 0 gm Fat • 7 gm Protein •
27 gm Carbohydrate • 437 mg Sodium •
230 mg Calcium • 1 gm Fiber

DIABETIC: 1 Skim Milk • 1 Fruit

Shamrock Shakes

I think food should be fun, don't you? I bet your kids will grin with delight when you serve them these "lucky" green drinks before you head out to watch the St. Paddy's Day parade . . . or even just a school soccer game. ☻ Serves 4 (1 cup)

1 cup Carnation Nonfat Dry Milk Powder
Sugar substitute to equal ¼ cup sugar
1½ cups water
1 teaspoon vanilla extract
5 to 6 drops green food coloring
2 cups sugar and fat-free vanilla ice cream
1 teaspoon mint extract
½ cup crushed ice

In a blender container, combine dry milk powder, sugar substitute, water, vanilla extract, and food coloring. Cover and process on HIGH for 15 to 20 seconds or until mixture is smooth. Add ice cream, mint extract, and crushed ice. Re-cover and continue processing on HIGH for 10 seconds or until mixture is thick and smooth. Serve at once.

Each serving equals:

HE: ¾ Skim Milk • ½ Slider • 16 Optional Calories

148 Calories • 0 gm Fat • 10 gm Protein •
27 gm Carbohydrate • 143 mg Sodium •
329 mg Calcium • 0 gm Fiber

DIABETIC: 1 Skim Milk • ½ Starch

Blueberries and Cream

If getting those glasses of milk your body needs is hard for you, I'm ready with creamy blender drinks that deliver the perfect "disguise." If you'd like this a little colder and thicker, try freezing the fresh blueberries on a tray for a little while before you blend it up.

☻ Serves 2 (1 cup)

1 recipe JO's Buttermilk
1 cup frozen unsweetened blueberries
½ teaspoon vanilla extract
Sugar substitute to equal 2 tablespoons sugar

In a blender container, combine **JO's Buttermilk**, blueberries, vanilla extract, and sugar substitute. Cover and process on BLEND for 20 seconds or until mixture is smooth. Serve at once.

Each serving equals:

HE: 1 Skim Milk • ⅔ Fruit • 6 Optional Calories

120 Calories • 0 gm Fat • 8 gm Protein •
22 gm Carbohydrate • 128 mg Sodium •
281 mg Calcium • 2 gm Fiber

DIABETIC: 1 Skim Milk • ½ Fruit

"Pink Lady"

Maybe you've tried this pretty drink at your favorite watering hole, but here's a quick and easy way to mix it up at home anytime. Isn't that color a dream?   ☻   Serves 1

1 recipe JO's Strawberry Milk Beverage Mix
1 cup water
½ teaspoon rum extract
½ cup crushed ice

In a blender container, combine dry **JO's Strawberry Milk Beverage Mix**, water, and rum extract. Cover and process on HIGH for 15 seconds. Add crushed ice. Re-cover and continue processing on HIGH for 20 seconds or until mixture is smooth. Pour into a tall stemmed glass and serve at once.

Each serving equals:

HE: 1 Skim Milk • 6 Optional Calories

80 Calories • 0 gm Fat • 8 gm Protein •
12 gm Carbohydrate • 123 mg Sodium •
276 mg Calcium • 0 gm Fiber

DIABETIC: 1 Skim Milk

Piña Colada Milk Drink

You're bound to feel as if you're on vacation when you enjoy this classic creamy blender treat! There's just something relaxing about pineapple, coconut, and creamy milk whipped into a thick shake.

○ Serves 2

1 recipe JO's Vanilla Milk Beverage Mix
1 cup (one 8-ounce can) crushed pineapple, packed in
*   fruit juice, undrained*
¾ cup water
1 teaspoon coconut extract
1 cup crushed ice

In a blender container, combine dry **JO's Vanilla Milk Beverage Mix**, undrained pineapple, water, coconut extract, and crushed ice. Cover and process on HIGH for 30 seconds or until mixture is smooth. Pour into 2 glasses and serve at once.

Each serving equals:

HE: 1 Fruit • ½ Skim Milk • 6 Optional Calories

120 Calories • 0 gm Fat • 4 gm Protein •
26 gm Carbohydrate • 63 mg Sodium •
155 mg Calcium • 1 gm Fiber

DIABETIC: 1 Fruit • ½ Skim Milk

Chocolate, Peanut Butter and Banana Shakes

How could any kid (of any age!) turn down a milkshake so full of favorite flavors? If you're looking for ways to increase your family's milk consumption, this is a super-duper way to win them over.

☻ Serves 2 (1 full cup)

2 recipes JO's Chocolate Milk Beverage Mix
1 cup cold water
1 cup (one medium) sliced banana
½ teaspoon vanilla extract
1 tablespoon Peter Pan reduced-fat peanut butter, creamy or chunky
½ cup crushed ice

In a blender container, combine dry **JO's Chocolate Milk Beverage Mix** and water. Cover and process on HIGH for 20 seconds or until mixture is smooth. Add banana, vanilla extract, peanut butter and crushed ice. Re-cover and process on LOW for 15 seconds or until mixture is smooth. Pour into 2 large glasses and serve at once.

HINT: Good topped with 1 tablespoon Cool Whip Lite, but don't forget to count the few additional calories.

Each serving equals:

HE: 1 Fruit • 1 Skim Milk • ½ Fat • ½ Protein

203 Calories • 3 gm Fat • 11 gm Protein •
33 gm Carbohydrate • 162 mg Sodium •
283 mg Calcium • 2 gm Fiber

DIABETIC: 1 Fruit • 1 Skim Milk • ½ Fat

Festive Menus for Family Occasions

"We Survived the Holidays" Winter Brunch

Baked Raisin-Pancake Squares
Grande Egg Skillet
Sausage Quiche Squares
Banana Cherry Coffee Cake

St. Patrick's Day Parade Potluck

Broccoli Farmstead Chowder
Green Beans Extraordinaire
Irish Cabbage Rolls
Banana Graham Cracker Dessert
Shamrock Shakes

Simple Summer Bridal Shower

Salmon Pasta Salad
Pink Cloud Salad
Jiffy Shrimp Curry
Impossible Zucchini Tomato Pie
Strawberries Romanoff Cheesecake Parfait
Raspberry Chocolate Jewels
Blueberries and Cream

Soccer Team Back-to-School Supper

Milwaukee Cheese-Veggie Soup
Corn Pudding
Layered Taco Salad
Hamburger Milk Gravy and Potatoes
Chocolate "Bon-Bon" Pie
Chocolate, Peanut Butter, and Banana Shakes

"Shop 'Til You Drop" Post-Thanksgiving Dinner

Homestyle Potato Soup
Southern Coleslaw
Sherry's Cranberry Salad
Wisconsin Turkey Sandwiches
Holiday Banana Nog Pie

"Make Your Resolutions Stick" New Year's Eve Party

South of the Border Shrimp Dip
Scalloped Corn and Carrots
Chicken Club Salad
"Candy Bar" Apple Salad
Yucatan Shepherd's Pie
Potato Dumplings with Roast Pork and Sauerkraut
Mercy Memories Pumpkin Pie
Banana Split Cheesecake

Making Healthy Exchanges Work for You

You're ready now to begin a wonderful journey to better health. In the preceding pages, you've discovered the remarkable variety of good food available to you when you begin eating the Healthy Exchanges way. You've stocked your pantry and learned many of my food preparation "secrets" that will point you on the way to delicious success.

But before I let you go, I'd like to share a few tips that I've learned while traveling toward healthier eating habits. It took me a long time to learn how to eat *smarter*. In fact, I'm still working on it. But I am getting better. For years, I could *inhale* a five-course meal in five minutes flat—and still make room for a second helping of dessert!

Now I follow certain signposts on the road that help me stay on the right path. I hope these ideas will help point you in the right direction as well.

1. Eat slowly so your brain has time to catch up with your tummy. Cut and chew each bite slowly. Try putting your fork down between bites. Stop eating as soon as you feel full. Crumple your napkin and throw it on top of your plate so you don't continue to eat when you are no longer hungry.

2. Smaller plates may help you feel more satisfied by your food portions *and* limit the amount you can put on the plate.

3. Watch portion size. If you are *truly* hungry, you can always add more food to your plate once you've finished your initial serving. But remember to count the additional food accordingly.

4. Always eat at your dining-room or kitchen table. You deserve better than nibbling from an open refrigerator or over the sink. Make an attractive place setting, even if you're eating alone. Feed your eyes as well as your stomach. By always eating at a table, you will become much more aware of your true food intake. For some reason, many of us conveniently "forget" the food we swallow while standing over the stove or munching in the car or on the run.

5. Avoid doing anything else while you are eating. If you read the paper or watch television while you eat, it's easy to consume too much food without realizing it, because you are concentrating on something else besides what you're eating. Then, when you look down at your plate and see that it's empty, you wonder where all the food went and why you still feel hungry.

Day by day, as you travel the path to good health, it will become easier to make the right choices, to eat *smarter*. But don't ever fool yourself into thinking that you'll be able to put your eating habits on cruise control and forget about them. Making a commitment to eat good healthy food and sticking to it takes some effort. But with all the good-tasting recipes in this Healthy Exchanges cookbook, just think how well you're going to eat—and enjoy it—from now on!

Healthy Lean Bon Appetit!

Index

Hawaiian Cottage Fruit Salad, 145
Healthy Exchanges
  cooking tips, 52–62
  for entire family, 35
  on fat, 49–50
  Four "Musts" of, 27–28
  goals of, 25–26
  for good health, 34
  ideas behind, 25
  ingredient measurement, 29–30
  low-fat recipes, 27, 33
  low-sugar recipes, 33–34
  for noncooking cook, 27
  nutritional information, 30–31
  origins of, 23–24
  portion control in, 30, 34
  on processed foods, 50–51
  real-people foods in, 26–27, 28, 59
  on sodium, 47–48
  ten commandments for success,
    36–38
  for weight loss, 24, 25–26, 32–33,
    39–46
Healthy Exchanges Food Newsletter, The,
  322–323
Heart disease, low-fat recipes and,
  33
Heartland Pork Stroganoff, 215
Hispanics, bone loss among, 3
Holiday Banana Nog Pie, 271
Homestyle Potato Soup, 88
Hypertension, 16
Hyperthyroidism, 4, 10
Hypoglycemia, low-sugar recipes and,
  33–34, 52

I
Ice cream
  brownie Alaska, 250–251
  shake, 300
  toppings, 58–59
Impossible Zucchini Tomato Pie, 151
Indigestion, calcium supplements and,
  13
Individual Brownie Alaska, 250–251
Ingredients
  advance preparation of, 36
  availability of, 28
  brands, 63–66
  homemade vs store-bought, 61–62

  measurement of, 29–30, 37, 53
  nutrient analysis of, 30–31, 69–70
  See also Substitutes
Intestinal malabsorption, 11
Irish Cabbage Rolls, 224–225
Iron, 49

J
Jamaican Banana Spice Cake with
  Coconut Topping, 266–267
Jaw, bone loss in, 16
Jiffy Shrimp Curry, 175
JO's Dairy Mixes, 69, 71–83
JO's Fettucchine Alfredo, 149

K
Kidney failure, 4
Kidney stones, 6
Kitchen timer, 37

L
Lactose intolerance, 6, 13, 20n
Lasagna
  asparagus ham casserole, 218–219
  macaroni, 164–165
  Pam's, 33, 202–203
Layered Taco Salad, 124–125
Leftovers
  freezing, 53
  rice and pasta, 54
Lemon
  desserts, 253, 254–255, 263
  yogurt, 60
Lemon Crunch Bars, 253
Lemon Pudding Tarts, 263
Leniwe Pierogi, 166–167
Linguine with Tuna & Veggies, 169
Lunch, calcium options for, 20
Lupus, 4

M
Macaroni
  and ham, 220
  taco casserole, 198–199
Macaroni Lasagna, 164–165
Magical Pumpkin Pie, 34, 276
Magic Chocolate Cake with Sauce,
  260–261
Magic Mocha Cheesecake, 282
Main dish recipes, 147–228

I want to hear from you . . .

Besides my family, the love of my life is creating "common folk" healthy recipes and solving everyday cooking questions in *The Healthy Exchanges Way.* Everyone who uses my recipes is considered part of the Healthy Exchanges Family, so please write to me if you have any questions, comments, or suggestions. I will do my best to answer. With your support, I'll continue to stir up even more recipes and cooking tips for the Family in the years to come.

Write to: JoAnna M. Lund
c/o Healthy Exchanges, Inc.
P.O. Box 124
DeWitt, IA 52742

If you prefer, you can fax me at 1-319-659-2126 or contact me via e-mail by writing to HealthyJo@aol.com.

If you're ever in the DeWitt, Iowa, area, stop in and visit me at "The House That Recipes Built" and dine at **JO's Kitchen Cafe**, "Grandma's Comfort Food Made Healthy!"

Ever since I began stirring up Healthy Exchanges recipes, I wanted every dish to be rich in flavor and lively in taste. As part of my pursuit of satisfying eating and healthy living for a lifetime, I decided to create my own line of spices.

JO's Spices are salt-, sugar-, wheat-, and MSG-free, and you can substitute them in any of the recipes calling for traditional spice mixes. If you're interested in hearing more about my special blends, please call Healthy Exchanges at 1-319-659-8234 for more information or to order. If you prefer, write to JO's Spices, c/o Healthy Exchanges, P.O. Box 124, DeWitt, IA 52742.

JO'S SPICES . . . A Healthy Way to Spice Up Your Life™

Now That You've Seen The Strong Bones Healthy Exchanges Cookbook, Why Not Order The Healthy Exchanges Food Newsletter?

If you enjoyed the recipes in this cookbook and would like to cook up even more of these "common folk" healthy dishes, you may want to subscribe to *The Healthy Exchanges Food Newsletter*.

This monthly 12-page newsletter contains 30-plus new recipes *every month* in such columns as:

- Reader Exchange
- Reader Requests
- Recipe Makeover
- Micro Corner
- Dinner for Two

- Crock Pot Luck
- Meatless Main Dishes
- Rise & Shine
- Our Small World

- Brown Bagging It
- Snack Attack
- Side Dishes
- Main Dishes
- Desserts

In addition to all the recipes, other regular features include:

- The Editor's Motivational Corner
- Dining Out Question & Answer
- Cooking Question & Answer
- New Product Alert
- Success Profiles of Winners in the Losing Game
- Exercise Advice from a Cardiac Rehab Specialist
- Nutrition Advice from a Registered Dietitian
- Positive Thought for the Month

Just as in this cookbook, all *Healthy Exchanges Food Newsletter* recipes are calculated in three distinct ways: 1) Weight Loss Choices, 2) Calories with Fat and Fiber Grams, and 3) Diabetic Exchanges.

The cost for a one-year (12-issue) subscription with a special Healthy Exchanges 3-ring binder to store the newsletters in is $28.50, or $22.50 without the binder. To order, simply complete the form and mail to us *or* call our toll-free number and pay with your VISA or MasterCard.

_____ Yes, I want to subscribe to *The Healthy Exchanges Food Newsletter.* $28.50 Yearly Subscription Cost with Storage Binder $_____

$22.50 Yearly Subscription Cost without Binder . . . . $_____

_____ Foreign orders please add $6.00 for money exchange and extra postage . . . . $_____

_____ I'm not sure, so please send me a sample copy at $2.50 . . . . $_____

Please make check payable to HEALTHY EXCHANGES or pay by VISA/MasterCard

CARD NUMBER: _____ EXPIRATION DATE: _____

SIGNATURE: _____
Signature required for all credit card orders.

Or Order Toll-Free, using your credit card, at 1-800-766-8961

NAME:_____

ADDRESS:_____

CITY: _____ STATE: _____ ZIP: _____

TELEPHONE:( ) _____

If additional orders for the newsletter are to be sent to an address other than the one listed above, please use a separate sheet and attach to this form.

MAIL TO: **HEALTHY EXCHANGES**
P.O. BOX 124
DeWitt, IA 52742-0124

1-800-766-8961 for customer orders
1-319-659-8234 for customer service

Thank you for your order, and for choosing to become a part of the Healthy Exchanges Family!

About the Author

JoAnna M. Lund, a graduate of the University of Western Illinois, worked as a commercial insurance underwriter for eighteen years before starting her own business, Healthy Exchanges, Inc., which publishes cookbooks, a monthly newsletter, motivational booklets, and inspirational audiotapes. Her first book, *Healthy Exchanges Cookbook*, has more than 450,000 copies in print. Her second book, *HELP: Healthy Exchanges Lifetime Plan*, was published in 1996. A popular speaker with hospitals, support groups for heart patients and diabetics, and service and volunteer organizations, she has appeared on QVC, on hundreds of regional television and radio shows, and has been featured in newspapers and magazines across the country.

The recipient of numerous business awards, JoAnna was an Iowa delegate to the national White House Conference on Small Business. She is a member of the International Association of Culinary Professionals, the Society for Nutrition Education, and other professional publishing and marketing associations. She lives with her husband, Clifford, in DeWitt, Iowa.

About the Introducer

Brian L. Levy, M.D., F.A.C.E., is a graduate of the Johns Hopkins University School of Medicine, a Fellow of the American College of Clinical Endocrinologists (F.A.C.E.), and on the faculty of the New York University School of Medicine, where he is an Assistant Professor of Clinical Medicine. He currently has a practice in Endocrinology and Metabolism in New York City.

MAI

12/23/97